SpringerBriefs in Oceanography

For further volumes:
http://www.springer.com/series/11754

Towhida Rashid

Holocene Sea-level Scenarios in Bangladesh

Springer

Towhida Rashid
Department of Geography and Environment
University of Dhaka
Dhaka
Bangladesh

ISSN 2196-1212 ISSN 2196-1220 (electronic)
ISBN 978-981-4560-98-6 ISBN 978-981-4560-99-3 (eBook)
DOI 10.1007/978-981-4560-99-3
Springer Singapore Heidelberg New York Dordrecht London

Library of Congress Control Number: 2013956526

Printed on acid-free paper

Springer is part of Springer Science+Business Media (www.springer.com)

Contents

Chapter 1
Introduction

1.1 Background of the Research

Sea level change is the product of many interrelated processes and an integrated measure of environmental phenomena, in terms of both causes and consequences. Changes occur on all space and time scales, from local (relative) to global (eustatic), and from a few second to geologic ages. Relative sea-level change is an important process influencing not only the relative position of the shoreline, but also the characteristics of coastal stratigraphic systems (Ta et al. 2001). It is noticeable that the stratigraphy of coastal lowlands is similar worldwide, being characterized by alternating strata composed of marine and terrestrial sediments (Tooley 1992). The strata occurring in a sedimentary succession are representative of different depositional systems that prevailed in the basin during its life term. In a sedimentary succession the products of the depositional systems are stacked vertically. The paleogeography of an ancient sedimentary basin at a particular time can be reconstructed from a sedimentary succession by knowing the set of depositional systems that existed in the space of the basin contemporaneously. As stated in Walter's law: from a package of sediments representing a continuum of sedimentation, the ancient array of depositional systems (i.e., the paleogeography) can be reconstructed by spreading out each depositional system (occurring vertically in the package) like a deck of cards. A clear understanding of the sedimentary successions reveal a precise sea-level change (transgression and regression) during the past, especially during the Holocene, which follows "Walther's Law of facies" (Walther 1924). According to the law, sedimentary environments that started out side-by-side will end up overlapping one another over time due to transgressions and regressions and the result is a vertical succession of beds (Levin 2006). It also states that the vertical succession of facies mirrors the original lateral distribution of sedimentary environments.

Although the interaction of eustatic and isostatic factors produces the general pattern of relative sea-level changes, various factors operate at coasts and within estuaries, influencing the registration of sea-level changes in the fossil and sedimentary records (Shennan and Horton 2002). In ancient successions, coastal zones

T. Rashid, *Holocene Sea-level Scenarios in Bangladesh*,
SpringerBriefs in Oceanography, DOI: 10.1007/978-981-4560-99-3_1,

are relatively easy to identify using fossils because not only may they contain a brackish water fauna but they also coincide with vertical and/or lateral changes from strata with dominantly marine fauna to strata with fresh water or no fauna. Hence landward migrations of the shoreline (transgressions) and seaward migrations of the shoreline (regressions) are comparatively easy to pin-point in both space and time as the rise of relative sea-level decreases the influence of terrestrial processes and increases the influence of marine processes (Islam 2001). As a result of transgression along with the rise of relative sea-level, fresh-water vegetation would be replaced by brackish-water mangrove species (Islam 2001). The composition (lithology) of coastal sediments, coupled with any biological components they contain, represents a rich source of information on past changes in relative sea-level (Godwin 1940; Tooley 1978).

During the last few decades considerable advances have been made in systematic sea-level investigation at different time scales. The methodology (Tooley 1978; Shennan 1982) is well established and needs to be applied over a wider range of geographical settings, particularly in the tropical coastal lowlands. The evidence for post-glacial sea-level movement can be found both above and below the present sea surface. Before the discovery of the C-14 dating technique by Libby in 1955, geomorphological features had been used to make the link between land-level and sea-level movement. Diatoms are also very useful to trace sea-level changes and have successfully been applied by Tooley (1978), Devoy (1979), Shennan (1986), and Ireland (1987) among others. Litho-bio-chrono-stratigraphic techniques to reconstruct the Holocene sea-level changes are well established and have been applied successfully in many coastal areas of the world including in Bangladesh (Tooley 1978; Shennan 1982; Ireland 1987; Islam 2001). Reconstruction of the Holocene sea-level history from geological evidence has successfully been applied to many coastal lowlands, such as the Netherlands (Jelgersma 1961), England (Tooley 1978), Shennan (1982), Brazil (Ireland 1987), China (Zong 1992) and many others. The World Bank has estimated that by the end of twenty-first century the relative sea level in Bangladesh could rise by as much as 1.8 m, and the increased landward intrusions of salt water would further damage water quality and agriculture including a huge loss of population (Maslin 2004). Moreover, Bangladesh lies less than 5 m above sea level and thus flooding is a common occurrence.

1.2 Rationality of the Research

The shoreline of Bay of Bengal has been observed not to be static in relation to previous geological events (Umitsu 1987, 1993; Kudrass et al. 1999; Goodbred and Kuehl 2000; Islam 2001). During the peak of the last glaciation (18000 years BP) the Bengal Lowland experienced dry climatic conditions and sea-level was 100 m or more lower than the present sea level (Umitsu 1987). At about

12000 years BP, south-west monsoon became prominent which caused heavy rainfall and sea-level started rising very rapidly (Monsur 1995). This amplified monsoon water plus deglaciated melt water from the Himalayas flowed over the Bengal Lowland and the initial Madhupur and Barind surfaces were highly dissected, creating some local pools and depressions (Monsur 1995). It seems that the coastal processes and climate had a significant control on topography of this region during Late Pleistocene time. The Indian Summer Monsoon (ISM) during early to middle Holocene was generally stronger than today, with peaks identified at 8500, 6400 and 2700 years BP detected in numerous ISM records (e.g., Fleitmann et al. 2003; Gasse et al. 1991; Van Campo et al. 1996; Wang et al. 2005), but weaker than today between 5000 and 1200 years BP. Umitsu (1993), Kudrass et al. (1999), and Goodbred and Kuehl (2000) stated that during the mid Holocene sea level of the bay was slightly higher, the climate was warmer, and rivers of this region discharged up to two and half times more than in present times. It has generally been accepted that around 6000 years BP, eustatic sea-level was higher than the present sea level. Morgan and McIntire (1959) told about six en-echelon faults on the western margin of the Madhupur Tract, which cause contrasting relief between Pleistocene and recent Holocene sediments, are noticeable. They told about en-echelon faults but unfortunately, neither field work nor examination of aerial photographs yielded evidence as to the type of movement along these faults. At present, based on some radiocarbon dating, it is thought that with some exceptions (where local block uplift is prominent), elevation of the Madhupur surface as a whole is quite apparent. Late Pleistocene monsoon climatic episodes caused a tremendous current flow over the Bengal plain and resulted dissection of the initial Madhupur surface. With the culmination of the amplified monsoon rainfall, the dissected valleys were filled up with the Holocene sediments created a geomorphic level at some lower position than the initial geomorphic surface (Monsur 1995).

As a coastal region, it was thought that there should be some evidences of Mid-Holocene marine transgression in and around Dhaka city. A limited number of detailed studies have been made in the last few decades to reconstruct the Holocene sea-level change of the Bengal Lowland, including those of Umitsu (1987, 1993), Banerjee and Sen (1987), and Islam (2001). These past attempts to reconstruct Holocene sea-level history have been based on borehole samples. The past studies did not determine the paleo-MSL and magnitude of the paleo-MSL, either. The past study areas are on low-lying deltaic deposits which are susceptible to long-term subsidence due to compaction of sediments by anthropogenic activities which may miscalculate the altitude of paleo-MSL. The present study considered six large outcrops for facies analysis. The study, the first of its type in the Bengal Lowland, presents a detailed description of the sedimentary facies, and discusses the relationship between these facies successions along with faunal analyses of pollen and diatom which yield the maximum information about the paleoenvironmental changes in this region. The study dealt with reconstruction of the paleoenvironment to know the relative sea-level history of Bangladesh during Holocene period. The present study also attempted to show the Holocene marine to

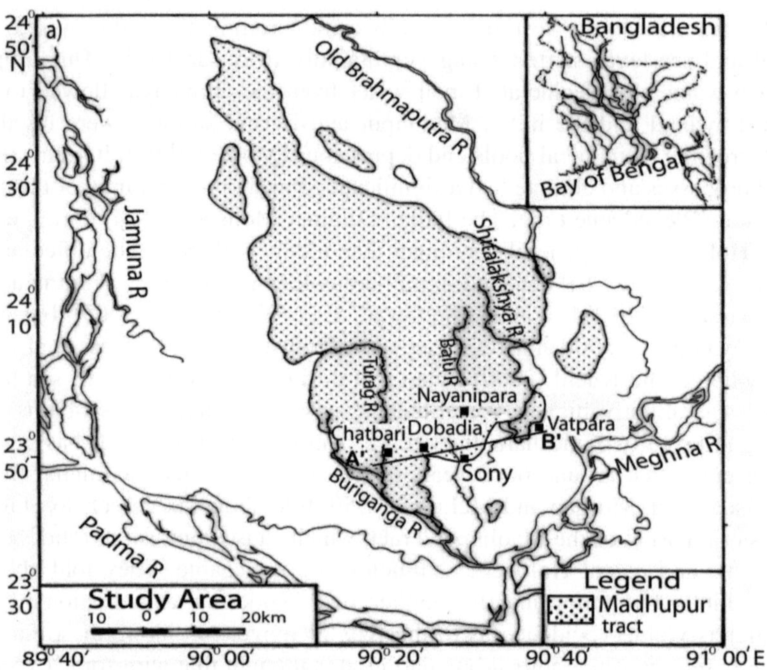

Fig. 1.1 The study area

brackish water invasion in and around Dhaka city and tried to discern the paleo-physiography in focusing to the evolution of the Madhupur Tract. With these aim, the study carried out lithofacies, biofacies (including pollen analysis) and diatom analyses, along with radiocarbon dating.

1.3 The Study Area

In order to reconstruct the paleoenvironment, to reveal the relative sea-level history of Bangladesh during Holocene period, Dhaka city has been selected as the study area. Dhaka is located in the central part of Bangladesh that is a large upland area called the Madhupur tract. The five outcrops at Sony, Vatpara, Chatbari, Dobadia and Nayanipara were selected from Narayanganj, Narshindi, Gazipur and Dhaka city (Fig. 1.1), respectively. These outcrops, in and around Dhaka city, were observed in eastern and south-eastern fringes of the Madhupur tract and are situated on the left bank of the Shitalakshya River as well as surrounded by numerous small rivers including Balu, Bangshi, and Turag (Fig. 1.2). The Madhupur tract is drained by the Dhaleshwari, Buriganga, Bangshi, Banar, and Shitalakshya of the Brahmaputra-Jamuna river system. All these rivers flow southeast and discharge

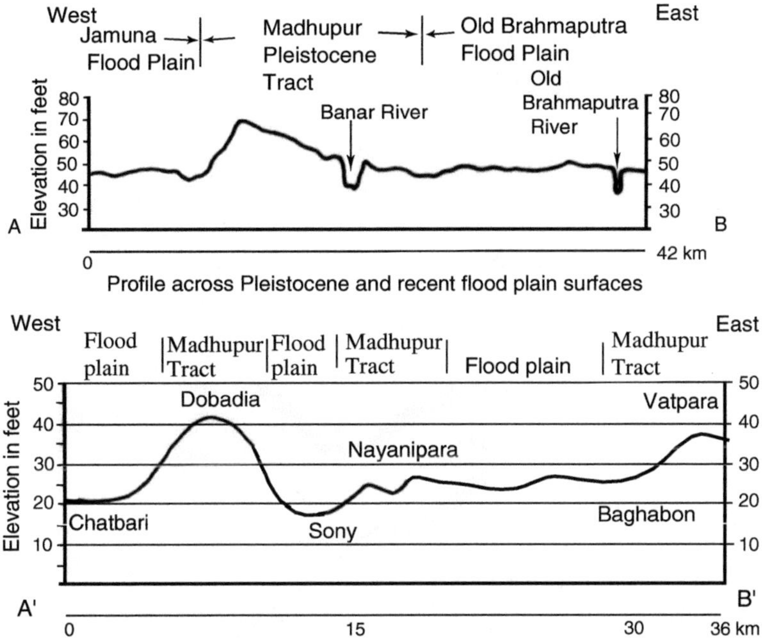

Fig. 1.2 Profile across the study area

Table 1.1 Description of the outcrops

Sl No.	Name of outcrop	Location	Physiographic description
1.	Sony	23°49' N, 90°29' E	It is situated on the left bank of the Balu River. It is an erosional dissected valley filled with Holocene sediments
2.	Vatpara	23°54' 59 N, 90°39'20 E	The site is situated on the left bank of Shitalakhaya River. The land is low-lying and undulating
3.	Chatbari	23°50'42" N, 90°20'33" E	Located on the left bank of the river Turag. It appears a deeply incised channel bed filled with Holocene sediments. The surface is flat. Dhaka International Airport is 5.8 km north-east from the Chatbari site
4.	Dobadia	23°51'45" N, 90°26'42" E	It is a deeply incised Madhupur tract valley filled with Holocene sediments. It appears to be the highest peak of the tract which is more susceptible to erosion. The surface physiography is undulating. The Dobadia is 10 km west from the site Sony and 5 km north from Dhaka Int'l Airport
5.	Nayanipara	23°56'15" N, 90°29'22" E	The surface physiography of the site is undulating. The site is situated on the left bank of river Balu

Fig. 1.3 Location of the study site Sony

freshwater into the vast river Meghna. The area annually inundates with flood water during monsoon time and after flooding it receives a small increment of fluvial sediments. The tract lies between the Jamuna and Old Brahmaputra Rivers. The study area, at present, is characterised by fresh water terrestrial vegetation including deciduous trees such as *gewa, sa*l etc. Table 1.1.

The study area is covered by two lithostratigraphic units, namely, Madhupur Clay Formation (Morgan and McIntire 1959) of the Pleistocene age and Basabo Silty-clay Formation of Holocene (Recent) age (Monsur and Paepe 1994). Upper part of Madhupur Formation is represented by highly weathered reddish-brown mottled clay, whereas the lower part is composed of micaceous sand having some

Fig. 1.4 Location of Vatpara

Fig. 1.5 Location of Chatbari site

Fig. 1.6 Location of Dobadia site

ripple marks and cross-bedded primary structures. The Madhupur Clay Formation is the oldest deposits exposed in the study area having erosional surfaces, floodplains, tract valleys and abandoned channels. Low lying swamps and marshes located in and around the city are other major topographic features. The incised channels and depressions in and around the area are floored by Holocene deposits. The floodplain deposits of Holocene age overlie the Madhupur Formation. The Holocene Series in the type locality at Bashabo is called Bashabo Formation (Monsur and Paepe 1994). The Bashabo Formation of Holocene age is represented by yellowish brown to bluish-grey sand to clay which is the youngest deposits widely occupied in the floodplain of the central Bangladesh. The five study sites and Bashabo are in the same locality, within 30 km, and lithology has been found similar. The Sony study area is only 14 km from Bashabo, and the lithology of both areas is similar. Hence, the deposits exposed at Sony are considered to be Bashabo Formation. Hence, the deposits exposed at five localities are considered as Bashabo Formation Figs. 1.3, 1.4, 1.5, 1.6 and 1.7.

Fig. 1.7 Location of Nayanipara site

References

Banerjee M, Sen PK (1987) Palaeobiology in understanding the changes of sea-level and coastline in Bengal Basin during Holocene period. Indian J Earth Sci 14:307–320

Devoy RJN (1979) Flandrian sea-level changes and vegetational history of the lower Thames Estuary. Phil Trans Royal Soc London 285:355–407

Fleitmann D, Burns SJ, Mudelsee M, Neff U, Kramers J, Mangini A, Matter A (2003) Holocene forcing of the Indian Monsoon recorded in a stalagmite from Southern Oman. Science 300:1737–1739

Gasse F, Arnold M, Fontes JC, Fort M, Gibert E, Huc A, Li BY, Lju Q, Melieres F, Vancampo E, Wang FB, Zhang QS (1991) A 13000-year climate record from western Tibat. Nature 353:742–745

Godwin H (1940) Studies of the post-glacial history of British vegetation, iii. Fenland pollen diagram, iv. Post glacial changes of relative land-and sea-level in the English Fenland. Phil Trans Royal Soc B230:239–303

Goodbred SL, Kuehl SA (2000) The significance of large sediment supply, active tectonism, and eustasy on margin sequence development: late Quaternary stratigraphy and evolution of the Ganges-Brahmaputra Delta. Sed Geol 133:227–248

Ireland S (1987) The Holocene sedimentary history of the coastal lagoon of Rio de Janiero state, Brazil. In: Tooley MJ, Shennan I (eds) Sea-level changes, Blackwell, Oxford

Islam MK (2001a) Evidences of neotectonics in and around Lalmai Hills, Bangladesh. Bangladesh J Geol 20:55–66

Islam MS (2001b) Sea-level changes in Bangladesh: the last 10000 years. Asiatic Aociety of Bangladesh, Dhaka

Jelgersma S (1961) Holocene sea-level changes in the Netherlands. Meded Geol Sticht Serie C 7:1–100

Kudrass HR, Spiess V, Michels M, Kottke B, Khan SR (1999) Transport processes, accumulation rates and a sediment budget for the submarine Delta of the Ganges-Brahmaputra, and the Swatch of No Ground, Bangladesh. In: International seminar of the Quaternary development and coastal hydrodynamics of the Ganges Delta in Bangladesh. Geol Soc Bangladesh, Dhaka, 20–21

Levin LH (2006) The earth through time. Wiley, New York

Maslin M (2004) Global warming: a very short introduction. Oxford University Press, New York

Monsur MH (1995) An introduction to the Quaternary geology of Bangladesh. City Press and Publications, Dhaka

Monsur MH, Paepe R (1994) Holocene stratigraphy and palaeoclimatological interpretation of the deposits of the Madhupur, Barind and Chalanbil areas of the Bengal Basin, Bangladesh. Bangladesh J Sci Res 12:255–264

Morgan JP, McIntire WG (1959) Quaternary geology of the Bengal Basin, East Pakistan and India. Geol Soc Am Bull 70:319–342

Shennan I (1982) Interpretation of Flandrian sea-level data from the Fenland. England Proc Geol Asso 83(1):53–63

Shennan, I (1986) Flandrian sea-level changes in the Fenland. II. Tendencies of sea-level movement, altitudinal changes and local and regional factors. Jour Quat Sci 1:155–179

Shennan I, Horton BP (2002) Holocene land and sea-level changes in Great Britain. J Quat Sci 17:511–526

Tooley MJ (1978) Sea-level changes during the last 9000 years in North-west England. Geog J 140:18–42

Ta TKO, Nguyen VL, Tateishi M, Kobayashi I, Saito Y (2001) Sedimentary facies, diatom and foraminifer assemblages in a late Pleistocene–Holocene incised-valley sequence from the Mekong River Delta, Bentre Province, Southern Vietnam: the BT2 core. J Asian Earth Sci 20:83–94

Tooley MJ (1992) Recent sea-level changes. In: Allen JRL, Pye K (ed) Saltmarsh; morphodynamics, conservation and engineering significance, Cambridge University Press, Cambridge, 19–40

Umitsu M (1987) Late Quaternary sedimentary environment and landform evolution in the Bengal Lowland. Geogr Review Jpn 60:164–178

Umitsu M (1993) Late Quaternary sedimentary environments and landforms in the Ganges Delta. Sed Geol 83:177–186

Van Campo E, Cour P, Sixuan H (1996) Holocene environmental changes in Bangong Co basin (western Tibet), part 2: the Pollen record. Palaeogeogr Palaeoclimatol Palaeoecol 120:49–63

Walther J (1924) Das Gesetz der Wuestenbildung in Gegenwart und Vorzeit, 4th edn. Quelle & Meyer, Leipzig

Wang YJ, Cheng H, Edwards RL, He YQ, Kong XG, An ZS, Wu JY, Kelly MJ, Dykoski CA, Li XD (2005) The Holocene Asian Monsoon: links to solar changes and North Altantic climate. Science 308:854–857

Zong Y (1992) Postglacial stratigraphy and sea-level changes in the Han River Delta. China J Coast Res 8(1):1–28

Chapter 2
Sea Level Research: Methods and Techniques

2.1 Introduction

Sea level can be recorded either from the direct instrumental measurements or from indirect geological evidence (Tooley 1992). Sea level change determination from geological evidence includes sedimentary records, diatom analysis, pollen analysis, carbon dating etc. On the other hand, direct observations of relative sea level changes have been derived mainly from tide gauge data. In the conventional tide gauge system sea level is measured relative to a land based tide-gauge benchmark (Emery 1980).

2.2 Sea Level Changes from Geological Records

2.2.1 Sea-Level Index Points

Over 30 years of international research, co-ordinated under the auspices of a series of International Geoscience Programme (IGCP) projects, has produced a well-defined methodology 'sea-level index points' and refined for developing records of relative sea-level change from sedimentary coasts (Edwards 2006). Central to this is the use of sea-level index points which fix the past altitude of sea level in time and space (Tooley 1978; Preuss 1979; Shennan et al. 1983; van de Plassche 1986). In brief, for a sample to be established as a sea-level index point, it must possess information regarding its location (latitude and longitude), its altitude (relative to a levelling datum), its age (commonly inferred from radiocarbon dating), and its vertical relationship to a contemporenous tide level (termed the indicative meaning). The later is important when accounting for the differing vertical distributions of coastal sub-environments and associated sea-level indicators. A sediment sample from a known location can be used as a sea-level index point if it is dated and its altitude is measured relative to a geodetic datum (Shennan 1982, 1986; Tooley 1982). Sea-level index points (Fig. 2.1) are commonly derived from

T. Rashid, *Holocene Sea-level Scenarios in Bangladesh*,
SpringerBriefs in Oceanography, DOI: 10.1007/978-981-4560-99-3_2,

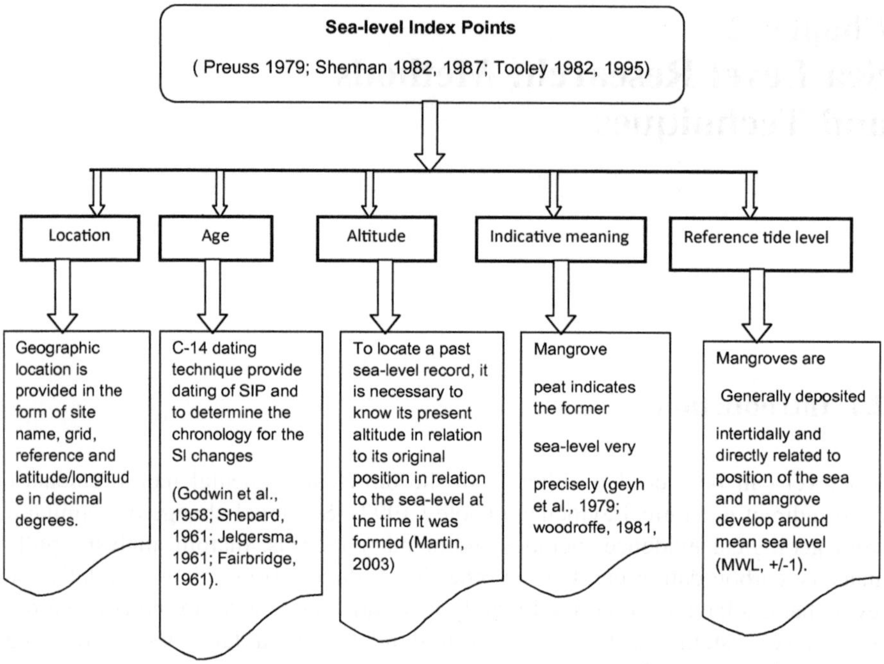

Fig. 2.1 Flow chart showing the five principal attributes of sea-level index points

lithostratigraphic contacts between terrestrial and marine sediments, with supporting microfossil data being used to delimit the onset or removal of brackish/ marine conditions. The elevation at which it formed relative to a contemporaneous tide level (termed the indicative meaning) must also be determined in order to account for the range of heights at which different coastal sedimentary environments are found (Van de Plassche 1986).

2.2.1.1 Geographic Location

This information can be provided in the form of site name, and latitude/longitude in decimal degrees.

2.2.1.2 Age

The discovery of the C-14 dating technique brought a revolution to sea-level research methodology. This technique has been utilized by a number of researchers (Godwin et al. 1958; Shepard 1961; Jelgersma 1961; Fairbridge 1961). Before the discovery of the C-14 dating technique by Libby in 1955, geomorphological features had been used to make the link between land-level and sea-

Table 2.1 Radiocarbon dates of the three sites in and around Dhaka city

Name of the Site	Altitude	Material	^{14}C Age Conventional	^{14}C Age Calibrated (cal BP)
Site-1(Sony)	+1.80	Peat	1760 ± 60	1820–1530
	+0.90	Peat	3560 ± 60	4080–4030
	−1.60	Peat	5819 ± 60	6670–6410
	−3.75	Wood	6690 ± 40	7570–7430
Site-2(Vatpara)	+0.40	Peat	4670 ± 60	5580–5300
Site-3(Nayanipara)	−0.20	Peat	3320 ± 50	3690–3440
Site-4 (Dobadia)	+5.7	Peat	5120 ± 50	5980–5970
	+4.7	Wood	5330 ± 60	6280–5940
Site-5 (Chatbari)	+3.1	Peat	2800 ± 70	3080–2760
	+2.1	Peat	4230 ± 60	4870–4780
	+0.9	Peat	4520 ± 70	5440–5410

level movement. The use of dates from different materials, without knowing their indicative meaning as to sea-level movements, may produce a completely false sea-level history at any location (Tooley 1978). Nine peat samples and two wood fragments were measured by the Radiometric and Accelerator Mass Spectrometry (AMS) method, respectively. 14C ages were determined by Beta Analytic Radiocarbon Dating Laboratory, Miami, Florida to explore the sea-level changes scenario of Bangladesh during the Holocene. For determining the age of the peat/ wood samples by radiocarbon dating, four samples were collected at Sony at altitudes of +1.80, +0.90, −1.60, and −3.75 m ASL, one was at Vatpara at +0.4 m and Nayanipara at −0.2 m, three were at Chatbari at +0.9, +3.1 and +2.1 m, two were at Dobadia at +5.7, 4.7 m, respectively (Table 2.1).

2.2.1.3 Altitude

A sediment sample from a known location can be used as a sea-level index point if it is dated and its altitude is measured relative to a geodetic datum (Shennan 1982, 1986; Tooley 1982). Datum level of any country is the reference level above or below which the altitude of any given point is measured (Jardine 1982). This is the most useful reference level for national surveying in any country. In Bangladesh, topographic maps are not yet been prepared on the basis of datum level as the datum levels are not uniformly pointed out in this country. To find the altitude at the top of the surface of six sections around in the central part of Bangladesh, the nearest datum levels of respective stations were used. Six leveling datum have been employed in research around the study region. With the help of the closest reference datum, Mean Sea Level (MSL) of the six outcrops was calculated by "differential leveling" method. Differential leveling (also called direct leveling) is used in determining elevations of points to establish a chain or network of BMs for future use. It requires a series of instrument setups along the survey route; and for setup, a horizontal line of sight is established, using a sensitive level.

2.2.1.4 Indicative Meaning and Reference Tide Level

The indicative meaning of a sample describes its vertical position relative to the tidal frame at the time of its formation. Where available, the database provides information on the inferred indicative meaning, the nature of the evidence from which this is derived, and the local tidal parameters used in reconstruction.

Sea-Level Indicators

Every marine feature or organism which has a quantifiable vertical relationship to a reference water level, the height of which is ultimately controlled by tidal amplitude at the coast, can be regarded as a sea-level indicator (Devoy 1987). On a global basis many different categories of indicators are used in sea-level studies, such as raised beaches, benchrock, notches, slope breaks shells, coral reefs, and archaeological features. Kidson (1981) urged that erosional features, such as shore platforms and notches can only rarely be used as precise indicators of former sea level.

Biological in depositional sequences

Any good sea-level must provide three essential elements: (1) accuracy, (2) preservability, and (3) be datable. Fossils within depositional sequence have the highest probability to provide all three of these elements.

Microfossil plant remains

The most useful indicators are microfossils because they occur in large numbers in small diameter cores which are often where the sea level records come from in Holocene submerged coastlines. Some attempts have been made to use pollen (Fig. 2.2) which is the microscopic part of the sexual reproductive organs of all angiosperm plants. Palynology is the study of pollen, spores, and similar palynomorphs, living and fossil. This is a powerful and useful tool to reconstruct the Quaternary environment (Birks and Birks 1980). The technique of qualitative pollen analysis was initiated by the Swedish geologist Lennart Von Post in 1916 (Birks and Birks 1980) which were later taken up and continued by his students (Faegri and Iversen 1989). Mangrove pollen is a very useful indicator of relative movements of sea-level in tropical environments (Belperio 1979; Ellison 1989), as well as the principal technique for reconstructing the Quaternary environment (Birks and Birks 1980).

A mangrove is a plant and *mangal* is a plant community and habitat where mangroves thrive. They are found in tropical and sub-tropical tidal areas, and as such have a high degree of salinity areas where *mangals* occur include estuaries and marine shorelines. Many, not all, mangrove swamps fringe estuaries and lagoons where the salinity changes with each tide. The tides rise along the coastline and enter the estuaries and travel inland as far as elevation allows.

Heritiera sp (m)

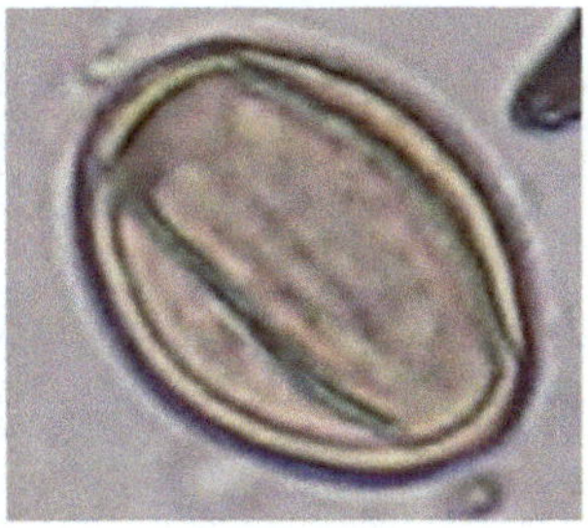

Aegicerus sp (m)

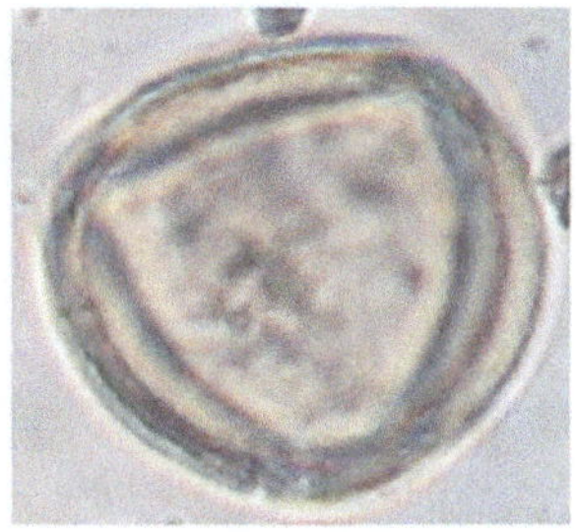

Xylicarpus sp(m)

Lumnitzera racemosa (m)

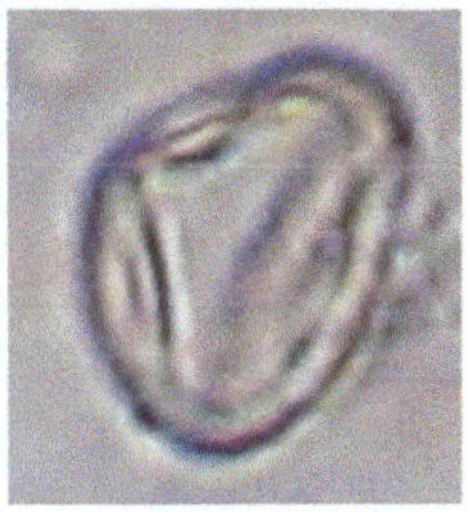

Bruguiera sp(m)

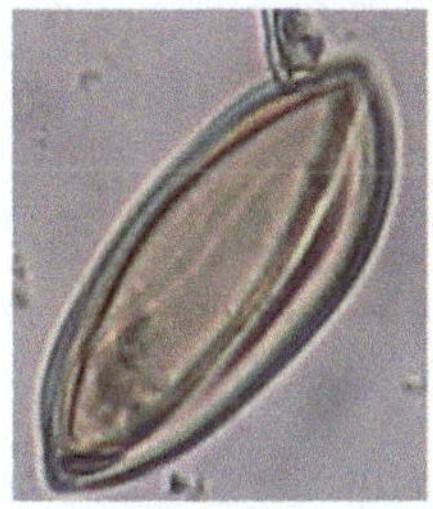

Phoenix paludosa (palm)

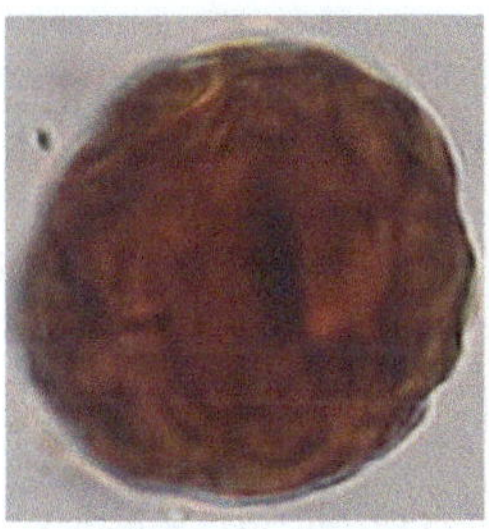

Temporina globate(a)

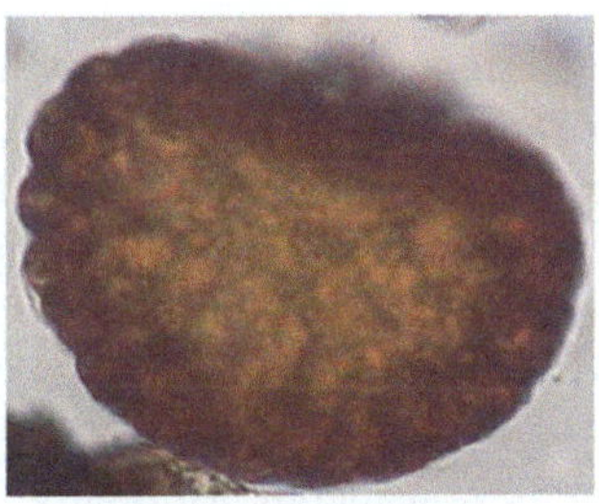

Corrugatisporites turpitus

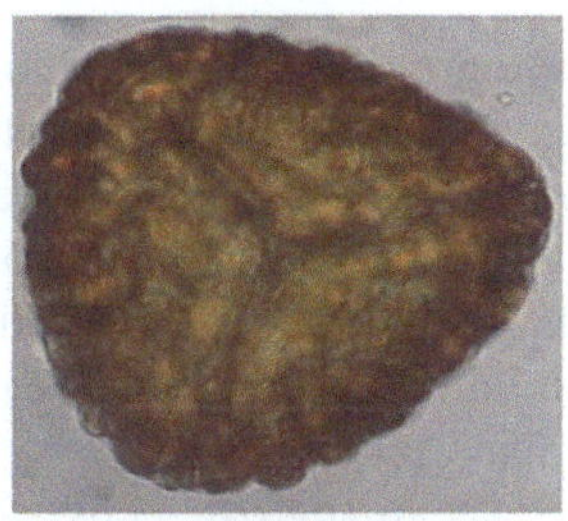

Tricolpites tressireticulatus

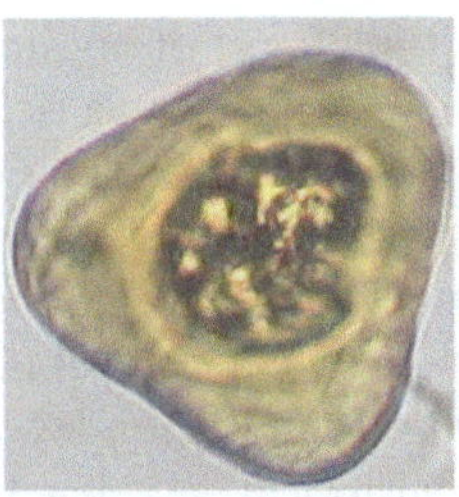

Dyadoporonites sp(f)

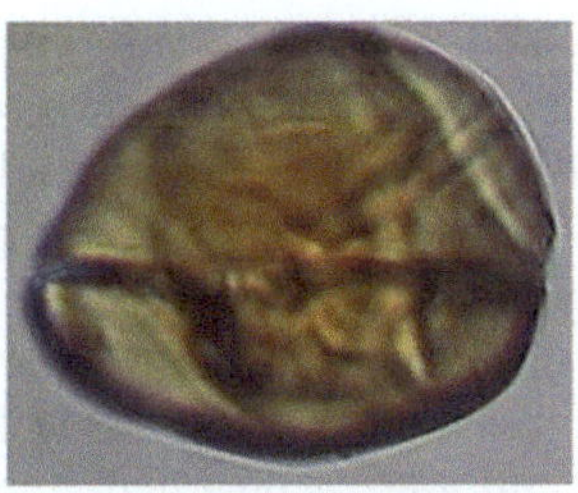

Dyadoporonites sp(f)

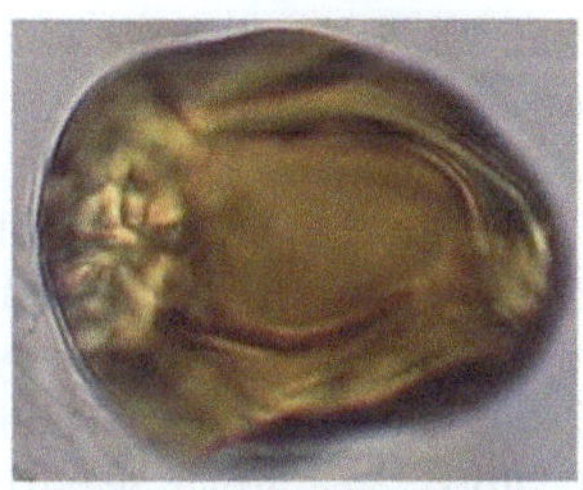

Octaplata palanaensis(a)

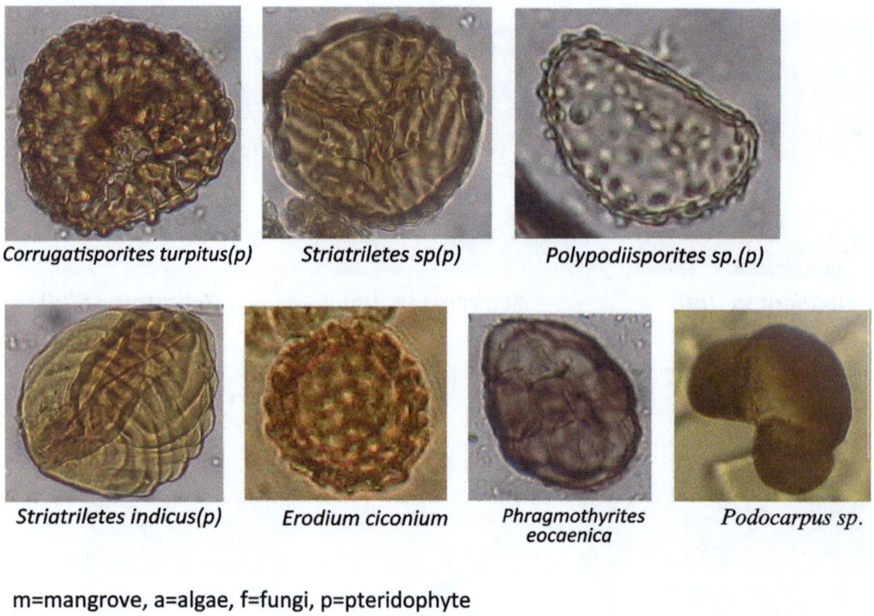

Fig. 2.2 continued

Mangrove swamp or *mangal* is one of an important habitat in brackish water environment. Brackish water is a result from mixing of seawater with fresh water with the salinity ranges between 0.5 and less than 35 %. Thus, rising sea level allows saltwater to penetrate farther inland and upstream (Gornitz 1991) and this effect impairs water supplies, ecosystems, and coastal farmland. In humid equatorial climates, gradual sea level rise may cause a brackish-water zone to migrate inland (Gornitz 1991).

The mangrove ecosystem is considered as highly susceptible to sea level changes (e.g. Gornitz 1991), and the sediments deposited beneath mangrove vegetation can provide useful indications of former sea levels (Scholl 1964; Woodroffe 1981; Van de Plassche 1986). The composition (lithology) of coastal sediments, coupled with any biological components they contain, represents a rich source of information on past changes in relative sea-level (Godwin 1940; Tooley 1978). Hence, the most reliable indicators are considered to be organic remains in their growth position, preferably in situ peat. Mangrove peats, which indicate the former sea-level very precisely, have been applied in some studies (e.g. Geyh et al. 1979; Woodroffe 1981; Woodroffe et al. 1985; Grindrod 1985, 1988) as sea-level indicators. Mangrove deposits aid the reconstruction of historical sea levels (Cohen et al. 2005).

Mangroves (generally) are trees and shrubs that grow in saline (brackish) coastal habitats in the tropics and subtropics. *Mangals* are found in depositional coastal environments where fine sediments, often with high organic content,

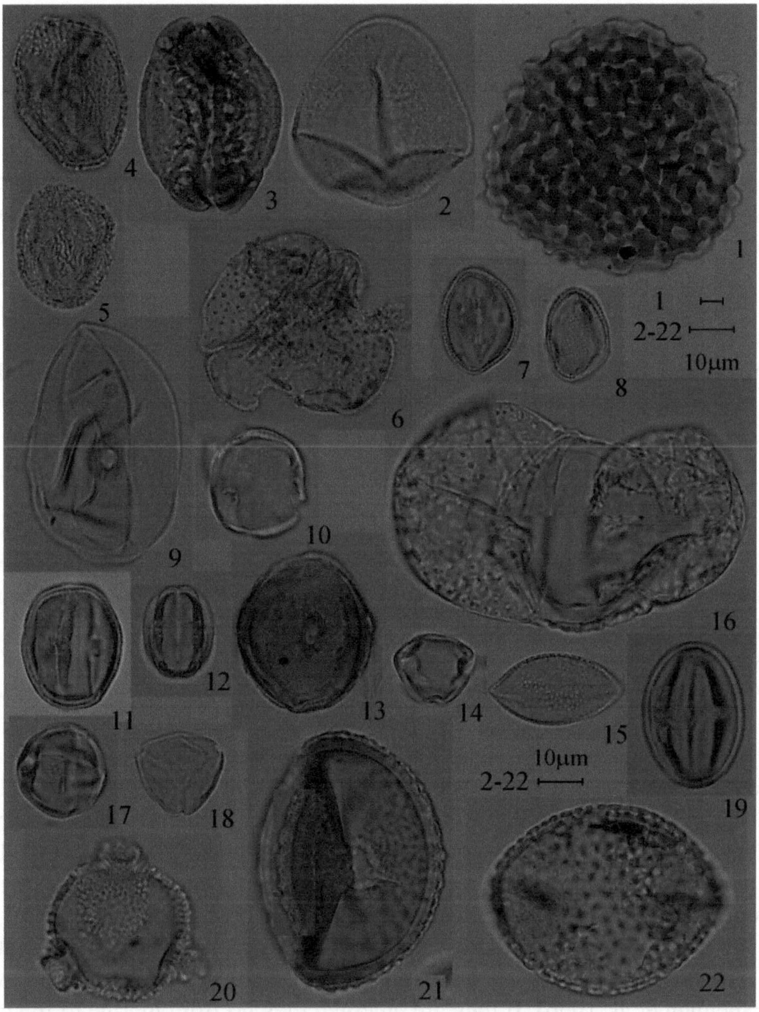

Fig. 2.2 Pollens and spores obtained in the study area. Pollen and spore types obtained from Sony with specimen number and sample horizon. *1*. Lygodium, NOS-9374, 1.8 m. *2*. Cyperaceae, NOS-9372, 1.8 m. *3*. Barringtonia, NOS-9369, 1.8 m. *4*. Semecarpus, NOS-9375. 1.8 m. *5*. Mangifera, NOS-9387, 1.2 m. *6*. Drosera, NOS-9386, 1.2 m. *7*. Phyllantus, NOS-9389, 1.2 m. *8*. unknown (tricolporate, prolate, small, fine-reticulate type), NOS-9492, 1.2 m. *9*. Gramineae, NOS-9394, 1.2 m. *10*. Vitis, NOS-9401, −1.6 m. *11*. Rhizophoraceae, NOS-9408, −1.6 m. *12*. Combretaceae. NOS-9424, −1.6 m. *13*. Lagerstroemia, NOS-9433, −1.6 m. *14*. Rhamnaceae, NOS-9412, −1.6 m. *15*. Arecaceae S type, NOS-9418, −1.6 m. *16*. Podocarpus, NOS-9406, −1.6 m. *17*. Rhizophoraceae, NOS-9448, −3.75 m. *18*. Myrtaceae, NOS-9441, −3.75 m. *19*. Leguminosae, NOS-9472, −3.75 m. *20*. Byttneria, NOS-9444, −3.75 m. *21*. Sonneratia, NOS-9452. −3.75 m. *22*. Arecaceae L type, NOS-9445. −3.75 m

collect in areas protected from high energy wave action. The geomorphic setting of mangrove systems comprises range of inter-related factors such as substrate types, coastal processes, sediment delivery, and freshwater delivery, all of which influence the occurrence of and survivorship of mangroves, while the tidal range will determine whether or not a sea-level rise will involve a significant proportion of the present mangrove habitat (Semeniuk 1994). Mangroves are generally deposited intertidally and may be directly related to the position of the sea at their time of deposition, and their sediments may contain large amounts of organic carbon suitable for radiocarbon dating (Woodroffe 1981). In reconstructing historical sea levels several studies (e.g. Scholl 1964; Geyh et al. 1979; Grindrod, 1985 and 1988; Mildenhall 1994; Clark and Guppy 1988; Bryant et al. 1992; Chappell 1993; Fujimoto and Miyagi 1993; Woodroffe et al. 1993; Beaman et al. 1994; Fujimoto and Miyagi 1993; Rashid et al. 2009, 2013) have departed from the assumption that mangroves develop around mean sea level. It is known that erosion, accretion or other disturbances may cause variation on the order of ±1 m (Bunt et al. 1985). Therefore, the topographical distribution of the mangroves in relation to mean sea-level is variable and may result in the sea level curve being displaced (Larcombe et al. 1995). In Grand Cayman (West Indies) the substrate of the marine mangrove is generally 15–30 cm above the mean sea level (Woodroffe 1981), and in Florida at 0–12 cm above the mean sea level (Scholl and Stuiver 1967). In the Townsville coastal plain (Australia), mangroves are found from 1.5 to 3.0 m amsl (Belperio 1979), while in Coral and Cocoa creek, northern Australia, the lower part of the mangrove forest is between 0 and 1 m above the mean sea level (Aucan and Ridd 2000).

Among all other microfossils, diatoms are also very useful to trace sea-level changes and have successfully been applied by Tooley (1978), Devoy (1979), Shennan (1986), Ireland (1987), Rashid (2009) among others. Diatoms are one-celled algae which leave a siliceous shell in the sediments (Shennan et al. 1998). These fossils appear to leave a record that can partially be subdivided the marsh into zones with varying accuracies (±20–50 cm at best). The most useful technique involving microfossil plants is determining sea levels in an indirect way where coastal ponds either become submerged or emerged (Palmer and Abbott 1986; Shennan et al. 1996). If they are emerged they go from marine to freshwater and vice versa for submergence; the key is to be able to determine the sill depth of the basin and then relate that to the radiocarbon dated transition. This method is usually accurate to within ±1 m depending on the depth and size of the basin (Laidler and Scott 1996). Either diatoms or dinoflagellates can be used in this manner. Dinoflagellates are also microscopic algae but they are organic-walled as opposed to the siliceous shells of diatoms. Both of these groups occur in large numbers such that 1 cm3 of wet sediment is often sufficient to obtain a valid result (Haq and Boersma 1978).

In case of diatom, they are useful for sea-level study because they are widespread in natural aquatic environments, many species prefer specific salinity conditions, the silica constituting the valves is relatively resistant to chemical alteration after burial and diatoms are often preserved within radiometrically dateable carbonaceous materials (Palmer and Abbott 1986).

Macro-animal fossils

Though the invertibrate macrofossils are limited in their value as sea level indicators by their vertical range in the water column, some groups such corals or attached biological indicators have extremely narrow ranges and have been used to produce some of the best and longest sea level curves (Fairbanks 1989). However, these are limited in occurrence and rocky intertidal forms are often not well preserved, especially in a submergent regime.

Corals may often be used as sea level indicator very effectively in tropical waters at depths ranging between MLWS and over 100 m. The depth varies in different coasts which are described below:

(a) On coasts with a strong vertical thermal gradient such as West Africa and Brazil, south of Cabo Frio; low water temperatures can be found at very shallow depths, and reef corals are thus confined to a very narrow vertical range (a few metres only). (b) On characteristic tropical coasts, where vertical thermal gradient is low, the composition of coral assemblage varies with depth just as the morphological and structural characteristics of the coral colonies do also. (c) Some reef coral species of the Indo-Pacific region are adapted to periodic emergence at low tide and are thus closely related to the low tide level. Some corals living in sheltered lagoons or in moats develop characteristic microatoll shapes: these may sometimes register water level variations.

The combination of these various criteria produced the best results, in the study of a fossil reef, when associated with morphological, sedimentological and geological studies. Special attention must be given to the possibility of coral colonies being broken by waves, displaced and later cemented once again on to the reef: large colonies of branching corals situated on the inner reef flat are generally good indicators.

Micro-animal fossils

There are two principal groups of animal microfossils that have been used extensively in sea-level studies, ostracods (Van Harten 1986) and foraminifera (Scott and Medioli 1986). Ostracodes leave a calcareous shell as a fossil and that in itself presents a problem because many of the best deposits for sea-level studies are not conducive to the preservation of CaCO3 hence the fossils are not present. Van Harten (1986) suggests the resolution with ostracodes to be within 100 m which not useful at all for modern se level studies but they can be used as accurate salinity traces and hence suggest transgression and regression (Haq and Boersma 1978).

On the other hand, foraminifera have both calcareous and agglutinated shells, agglutinated shells are resistant to dissolution in low pH, highly organic sediments, and often are very abundant in some highly organic deposits such as marshes (Scott and Medioli 1980).

Beachrock: modern indicator

In the catalogue of sea-level indicators, beachrock plays a significant role because of its wide distribution and extension on long stretches of coastlines.

Beachrock is defined by Scoffin and Stoddart (1987) as "the consolidated deposit that results from lithification by calcium carbonate of sediment in the intertidal and spray zones of mainly tropical coasts." Beachrock units form under a thin cover of sediment and generally overlie unconsolidated sand, although they may rest on any type of foundation. Maximum rates of subsurface beachrock cementation are thought to occur in the area of the beach that experiences the most wetting and drying-below the foreshore in the area of water table excursion between the neap low and high tide levels (Amieux et al. 1989; Higgins 1994). The majority of recent beachrock is formed on beaches in the same regions that favour coral reef formation (Turner 2005). Most publications, however, deal with beachrock cements rather than their incorporation into the coastal environment and its evolution. The relic character and age of most beachrocks has been overlooked, and beachrocks are called "modern" because of their position at or close to sea level.

Indicators of abandoned shorelines may be categorized as geological, biological, or archeological. Relict abrasion and accumulative relief types such as abrasion terraces, wave-cut notches, beach ridges, and aggradational terraces are related to former sea levels and can provide useful data for these reconstructions. Erosional scarps and coastal solution holes can also be useful indicators of former sea stands. In tropical and sub-tropical areas, beach-rock formation and deposits of oolite are often useful indicators of abandoned shorelines, as are peats and marsh and mudflat deposits. Organic materials such as peat and mollusk shells can be dated by radiometric methods.

2.2.1.5 Sedimentary Facies

Sedimentary rocks can be formed only where sediments are deposited long enough to become compacted and cemented into hard beds or strata. Sedimentation commonly occurs in areas where the sediment lies undisturbed for many years in sedimentary basins. Whereas some such basins are small, others occupy thousands of square kilometres and usually have within them several different local depositional environments. Physical, chemical, and biological factors influence these environments, and the conditions that they produce largely determine the nature of the sediments that accumulate. Several different local (sedimentary) environments may thus exist side by side within a basin as conditions change laterally; the sedimentary rocks that ultimately are produced there can be related to these depositional environments. These different but contemporaneous and juxtaposed sedimentary rocks are known as sedimentary facies.

In geology, a **facies** is a body of rock with specified characteristics. Ideally, a facies is a distinctive rock unit that forms under certain conditions of sedimentation, reflecting a particular process or environment (Encyclopedia Britanica online 2012). The term facies was introduced by the Swiss geologist Amanz Gressly in 1838. Facies is the study and interpretation of the textures, sedimentary structures, fossils, and lithologic associations of the sedimentary successions (Miall 1990) to interpret a paleoenvironment. It is a body of rock with specified

Table 2.2 Description of bedding thickness

	Thickness (cm)	Descriptive term
1	100+	Very thick bed
2	30–100	Thick bed
3	10–30	Medium bed
4	3–10	Thin bed
5	1–3	Very thin bed
6	0–1	Laminae: Thick laminae, 3–10 Thin laminae, 0–3

characteristics where sedimentary rocks can be handled on the basis of color, bedding, composition, texture, fossils and sedimentary structures. A biofacies is one for which prime consideration is given to the biological content. If fossils are absent or of little consequence and emphasis is on the physical and chemical characteristics of the rock, then the term lithofacies is appropriate.

Sedimentary structure

Sedimentary structure is the features formed in sediment at time (or very shortly thereafter) of deposition. Bedding/Stratification is most characteristic feature of sedimentary deposits. Fluctuation in current causes preferential deposition of grains. In some instance currents impart a distinctive orientation to deposited grains. In other cases grains on sediment may become packed differently during or shortly after deposition. Any of these factors can cause sediment to have a different appearance from that above and below, and therefore be observed as a separate bed. Bedding is generally defined as a sedimentary layering thicker than 1 cm (Table 2.2). The finer scale layering, only millimeters thick, is termed lamination (Table 2.2). Lamination is frequently an intertidal structure of a bed. The majority of beds have been deposited over a period of time ranging from hours or days, as in the case of turbidities and storm beds, 2 years, ten of years or even larger, as in the case of many marine shelf sandstones and limestones. The surface separating one bed from another is known as the bedding plane.

(A) Massive beds–beds with very little or no internal structures (Table 2.2; Fig. 2.3) which deposited under single uniform flow conditions or internal structures subsequently destroyed by bioturbation (disturbance by organisms).
(B) Graded bed—bed shows a progressive vertical change in grain-size—normal grading (fines upward) is common, but reverse grading (coarsen upward is rare).
(C) Cross-bedding—inclined layering commonly formed by deposition on a sloping face such as the inclined surfaces of bars, dunes, and sand waves.
(1) Planar cross-bedding—cross-beds parallel one another and intersect bedding at a consistent angle– formed by sand waves (straight crested dunes)

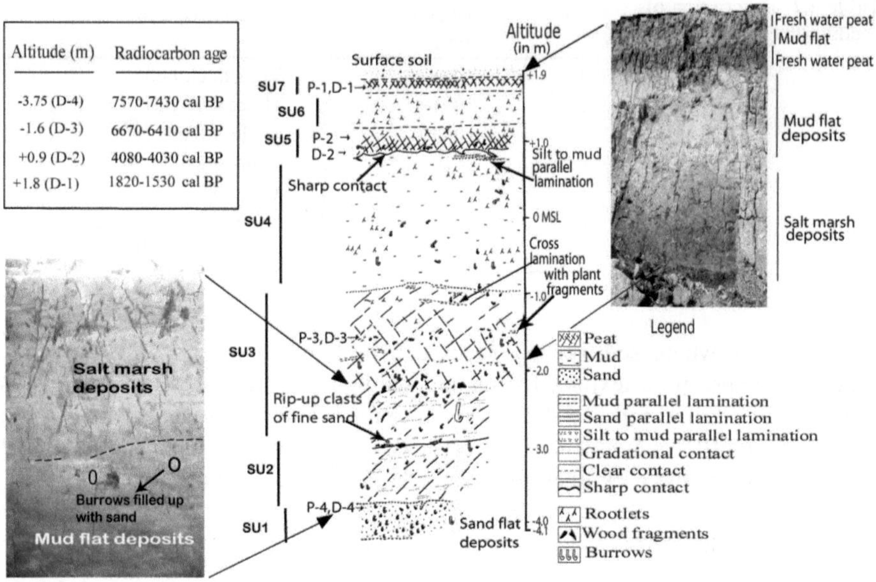

Fig. 2.3 Lithostratigraphic description at Sony

(2) Trough cross-bedding—cross-beds non-parallel and approach one another near base of bed becoming tangential or asymptotic to base– formed by sinuously crested dunes.

(D) (Ripple) Cross-lamination—current similar to cross-bedding (Fig. 2.3), but sets less than 3 cm thick, ripple have asymmetric form (one side steeper), and can have straight or sinuous crest or be isolated, formed unidirectional flow.

Wave ripples formed by bidirectional flow. Ripple have symmetrical form, sharp crest form in lower energy settings and rounded crest in higher energy.

Litho-bio facies

There are several ways of describing or designating sedimentary facies. By noting the prime physical (or lithological) characteristics, one is able to recognize lithofacies. The biological (or more correctly, paleontological) attributes—the fossils—define biofacies. Both are the direct result of the depositional history of a basin. By ascribing modes of origin to different facies (i.e., interpreting the lithofacies or biofacies) one can visualize a genetic system of facies. It is also common to speak of alluvial facies, bar facies, or reef facies, using the environment as a criterion.

Walther's Law of Facies, named after the geologist Johannes Walther, states that the vertical succession of facies reflects lateral changes in environment. Conversely, it states that when a depositional environment "migrates" laterally, sediments of one depositional environment come to lie on top of another. A classic example of this law is the vertical stratigraphic succession that typifies marine trangressions and regressions.

From studies of facies relationships to one another it has become recognized that the gradational, sharp, or eroded contacts between these rock bodies are also of significance in finding the mode of origin. Non-erosional contacts include that the facies immediately followed each other in time, probably by the migration of depositional environments. If contacts are sharp, even when erosion cannot be demonstrated, the facies may have been formed in depositional environments. Some contacts show extensive boring, burrowing, penecontemporaneous deformation or diagnosis of the underlying sediments so that the adjacent facies have become mixed or even inverted.

Today it is recognized that facies associations and distribution depend upon interrelated controls. The most important include sedimentary processes, sediment supply, climate, tectonics (earth movements), sea level changes, biological activity, water chemistry, and volcanic activity. Of these the environment of deposition (climate) and tectonic activity are paramount as they may ultimately regulate the other factors.

Facies association

Facies associations are groups of facies that occur together and are considered to be genetically or environmentally related. Each grouping would be identified as a distinct association. The association provides additional evidence which make environmental interpretation easier than treating each facies in isolation.

The litho- and biofacies were carefully observed to explore the paleoenvironments of the Holocene deposits at each site. On the basis of facies, together with the faunal analysis of pollen and diatom, the Holocene deposits around the study area (the central part of Bangladesh) have been assigned broadly into two sedimentary facies association; tidal environment facies association and fluvial facies association.

Tidal environment facies association

This facies association has been subdivided into 4 facies. Facies A: fine-grained sand; Facies B: marine mud; Facies C: marine organic mud; and Facies D: mud with interlaminated fine-grained sand or organic mud. These facies are described and interpreted bellow.

Fine-grained sand facies

The Facies is a massive muddy fine-grained sand to silt. Abundant burrows with sand pipes are developed at Sony (Fig. 2.3). The diameter of the pipes is approximately 1 cm. The pollen assemblages (Rashid et al. 2009) showed abundant occurrences of mangroves, such as Rhizophoraceae (28.9 %), *Sonneratia* (3.5 %), and *Avicennia* (0.9 %), with abundant occurrences of tropical forest elements. Dense burrows indicate high biogenic activity, which supports a tidal environment at Sony. Abundant mangrove pollens indicate that the mangrove ecosystem was developed very close to the deposited site. Pollens from tropical

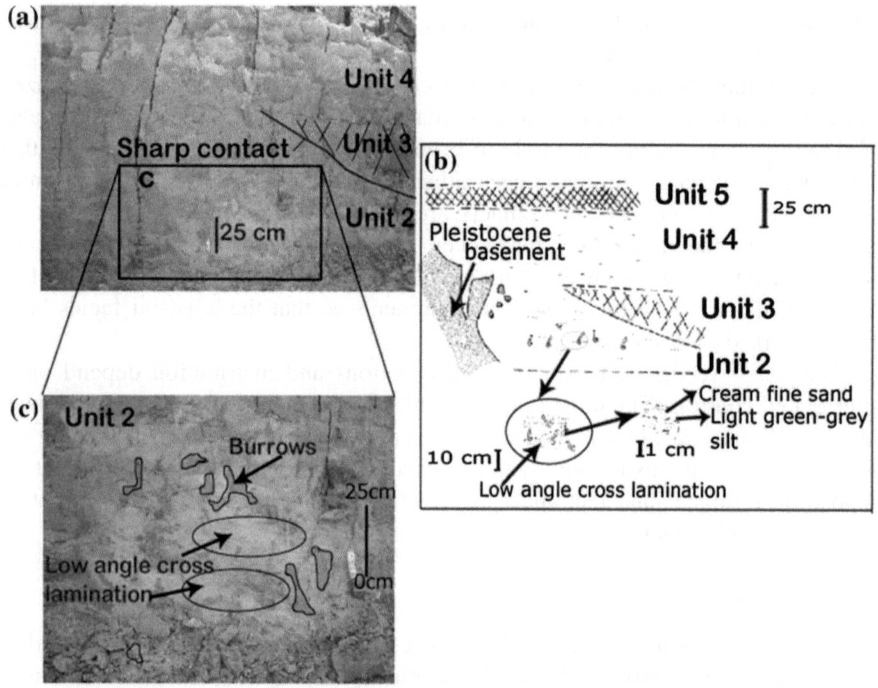

A. Photograph showing Unit 2 and Unit 3 at the southern wall of the outcrop.
B. Sketch of photograph A. Low angle cross lamination, burrows, cream-fine sand are shown in the sketch. Sharp contact underlying with mud flat are shown.
C. Sedimentary structures of Unit 2 (extracted from photograph A).
The photograph shows low angle cross laminations and big burrows filled with fine sand.

Fig. 2.4 Detailed descriptions of litho-facies at Sony

forests imply that the mangrove ecosystem had some fresh water influence. Considered with the sandy lithofacies, the environment represents a sand flat of an intertidal zone near the low water/tide line in a coastal-plain (Rashid et al. 2009).

Marine mud facies

The lithology of this facies is characterised by mud (Figs. 2.3 and 2.4). It contains a significant number of large burrows filled with fine cream-coloured sands. Rootlets and small burrows are observed here. Parallel laminations and small scale low angled cross laminations of mud and silty mud or very fine-grained sand are preserved in some parts of this facies. The Vatpara site contains some organic matter. Wood stem fragments and a large quantity of leaf fragments are found at Dobadia (Fig. 2.5). There are many pipes, occupied by organisms, in the wood stems, which are filled with mud sediments. Abundant mangrove pollens such as *Lumnitzera racemosa*, and *Xylicarpus* sp. are detected in the mud sediments. Also, a significant number of marine diatom fossils such as *Cyclotella stylorum*, *Cyclotella striata*, *Paralia sulcata*, *Tryblioptychus coccconeiformis*,

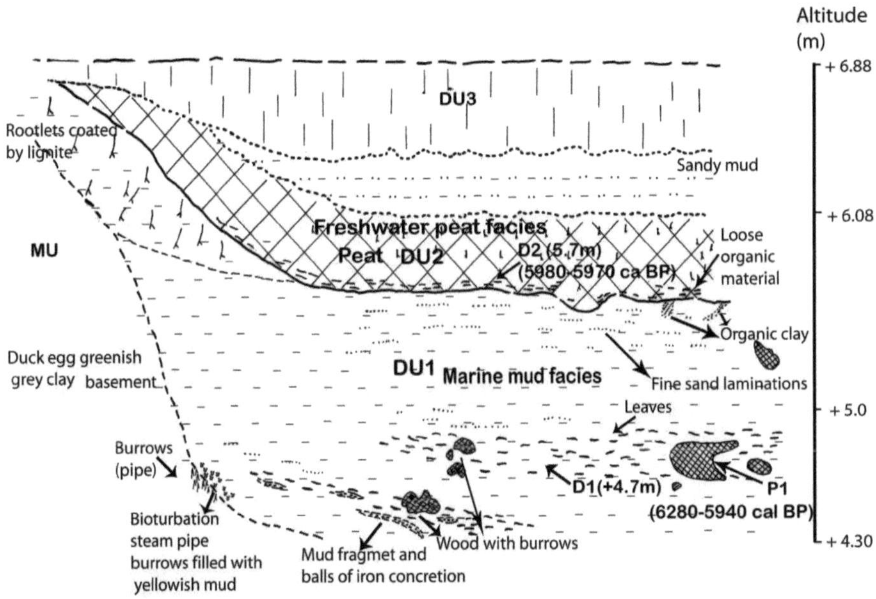

Fig. 2.5 lithostratigraphic description at Dobadia

Coscinodiscus radiatus, *Giffenia cocconeiformis, Coloneis westii,* and *Thalassiothrix* sp. are extracted from the sample from Dobadia. *Aulaeoseira granulata* is the only freshwater planktonic diatom found in Dobadia (Rashid et al 2013). The muddy lithofacies with parallel lamination and small-scale low angle cross lamination can be interpreted as a low-energy depositional environment at Dobadia. Marine evidence from the diatom fossils and mangrove pollens suggests that the mud was deposited in an intertidal mud flat (Rashid et al. 2009). The bioturbation also supports the mud flat environment. The lamination with fine-grained sand is thought to be formed by a tidal current. The sediment of this facies is interpreted as mud flat deposits.

Marine organic mud facies

Peaty mud with a high content of wood and plant fragments is the significant characteristics of this facies. Abundant small and large burrows are observed in Sony. Rootlets are common in the Vatpara and Sony. The pollen assemblage contains remarkable occurrences of Rhizophora (%), *Sonneratia* (%) with *Nypa* (%) and abundant algae and fungal and pteridophytic spores. Marin to brackish water diatoms are also obtained from the Sony (Rashid et al. 2009). The Vatpara is also characterised by mangrove pollens such as *Rhizophora* sp., *Bruguiera* sp, *Aegicerus* sp., *Heritiera* sp., *Xylicarpus* sp., and *Phoenix paludosa* (Rashid et al 2013). Some cross laminations of plant fragments were observed in Sony (Rashid et al. 2009). Thin layers of fine-grained sand are interlaminated in Sony and Vatpara (Figs. 2.3 and 2.6). The bottom of SU3 cuts the lower unit. Part of the boundary is covered by rip-up clasts of fine-grained sand fragments (Rashid et al.

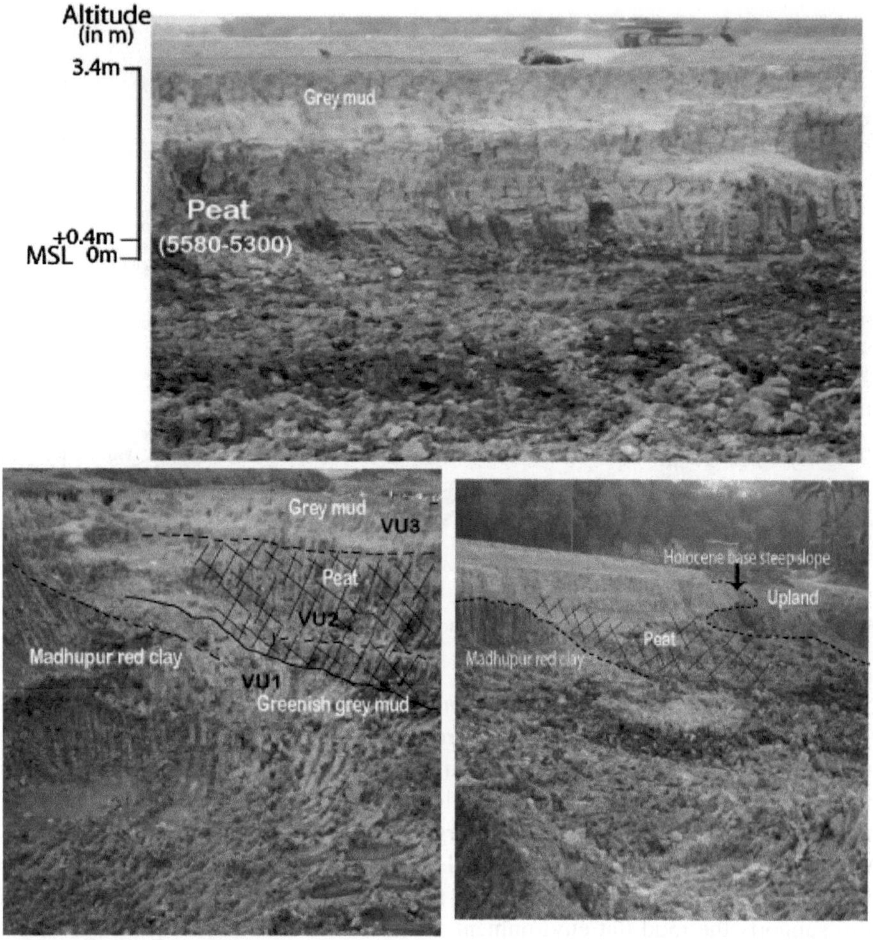

Fig. 2.6 Stratigraphic description at Vatpara

2009). The bottom of VU2 has a distinct, sharp boundary. Organic rich muddy lithofacies indicate the deposition of a marshy environment. Abundant occurrences of mangrove pollens and marine to brackish water diatoms suggest a salt marsh environment. The occurrence of rootlets supports the marshy environment. The abundant algae, spores and interlamination of very fine-grained sand reflect that the area was not permanently submerged and that it used to subject to a tidal effect.

Mud with interlaminated fine-grained sand or organic mud facies

A parallel lamination of fine-grained sand is developed in Chatbari site (Figs. 2.7, 2.8). It comprises an alternation of mud and organic mud. The mud beds are weakly laminated and a few thin (approximately 1 cm) sand beds are associated with them. The organic mud beds are composed of abundant plant fragments (including wood fragments) and mud. Channel structures are recognised at

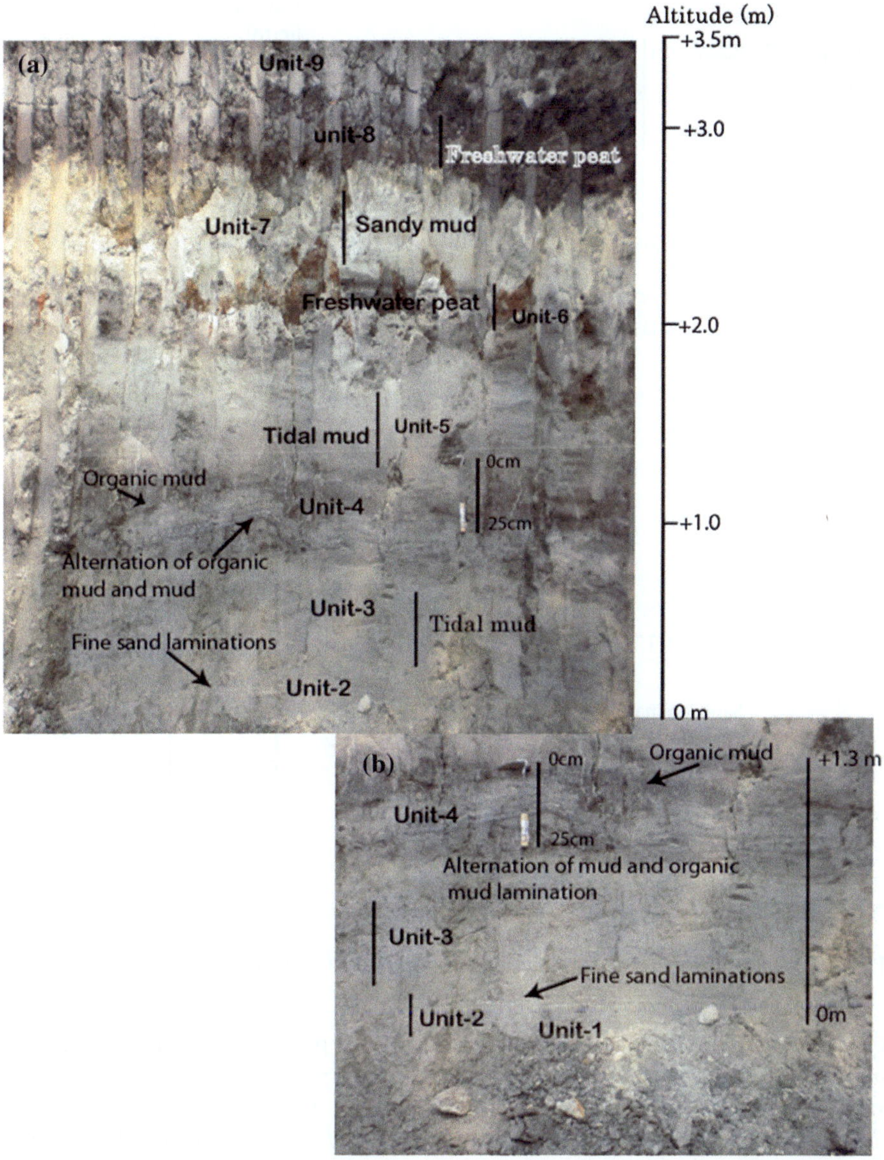

Fig. 2.7 Description of lithostratigraphy at Chatbari

the bottom of a few organic mud beds, which cut the lower bed. This facies is characterised by an occurrence of abundant diatom fossils. Under the microscope, fine-grained sand to silt clasts of quartz, mica (mainly biotite) with minor amounts of feldspar are observed in Chatbari (Rashid et al 2013). Feldspar clasts are not weathered. Soil structures such as sepic plasmic fabric are not developed in the

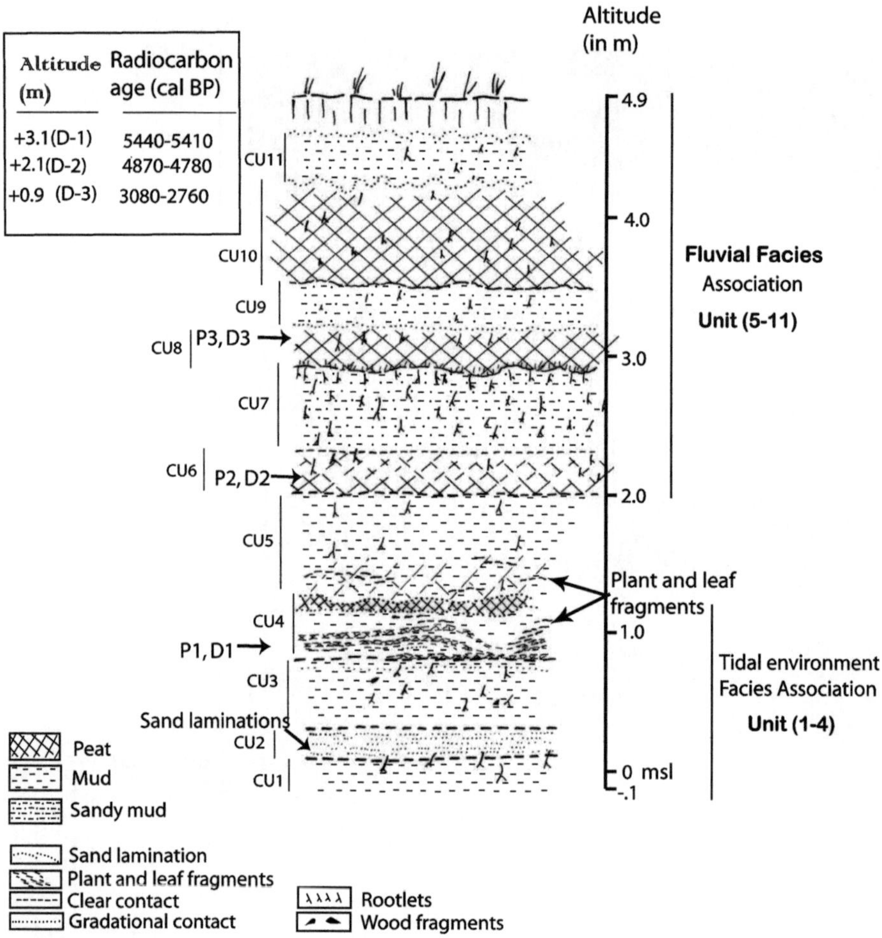

Fig. 2.8 Lithostratigraphic and sedimentary description at Chatbari

clay portion. This facies contains a few mangrove palm pollens, such as *Phoenix paludosa*. Abundant diatom fossils, such as marine planktonic diatoms (*Cyclotella striata*, *Actinocyclus octonarius* and 5 species), brackish planktonic diatoms (*Thalassiosira bramaputrae*, *Actinocyclus normanii* and 1 species) and freshwater planktonic diatoms (*Aulacoseira ambigua*, *A. granulata* and 3 species), are observed (Rashid et al 2013). A small number of haptobenthic diatom fossils, but with many species (3 species of marine haptobenthic diatoms, seven species of brackish haptobenthic diatoms and 21 species of freshwater haptobenthic diatoms), were also collected. It yields abundant freshwater diatoms, such as *Aulacoseira ambigua*, *Eunotia formica*, *Aulacoseira granulata*, *Aulacoseira granulate,* with two species of marine diatoms, such as *Cyclotella striata* and *Cyclotella stylorum* (Rashid et al 2013).

The general lithology of mud (silt with clay) with fine sand beds and organic mud indicates a low energy environment. The associated mangrove pollen and diatom fossils reveal a brackish water environment. The alternation of mud and organic mud beds indicates that the succession at Chatbari might have been affected by repeated flooding or tidal currents at the final stage of the deposition of this facies. The pollen fossils indicate that the organic matter was transported from nearby mangrove vegetation coverage, which might have been slightly more elevated than the deposition site. The Chatbari site is located in the valley of the Turag River and is situated at the border of the Madhupur Tract and a recent fluvial plain.

Considering the geographical setting, this facies is interpreted as the deposition of an estuarine environment. The predominance of the planktonic diatoms (marine, brackish and freshwater) indicates a lagoonal environment. The central basin of an estuary (Dalrymple et al. 1992) is thought to be reconstructed for a sedimentary environment.

Fluvial facies association

The fluvial facies association is subdivided into facies E: non-marine mud and facies F: peat, which are described and interpreted below.

Non-marine mud facies

The structureless ill-sorted sandy mud with abundant rootlets characterises the facies. Under a microscope, the pores of rootlets, coated with orange to deep red iron oxides, and clay skin are observed. In the mud matrix, clinobimasepic microfabrics (Retallack 1997) with highly birefringent streaks are developed.

This facies indicates that the sediments were formed by a mud flow. Abundant rootlets and soil structures reveal that the deposits were affected under a subaerial environment. The facies is interpreted to be deposited by flooding in a fluvial plain.

Peat facies

The facies is characterised by black massive peat, which is composed of mud and a large quantity of loose and undecomposed organic materials with small rootlets (Fig. 2.9). Abundant terrestrial plant pollen fossils are contained. The palynological data give evidence of freshwater sequences at Sony (Rashid et al. 2009). The pollen fossils from Dobadia are dominated by terrestrial tree elements such as Podocarpaceae and Geraniaceae. Podocarpaceae indicates a forest in moist conditions (Rashid et al 2013). DU2 is also characterised by pteridophytic and fungal spores. CU6, CU8, CU10 (Chatbari) and NU2 and NU4 (Nayamipara) contain terrestrial pollens and some pteridophytic and fungal spores. Under the microscopic observation of CU6 and CU8, some rootlet pores were coated with iron oxide. The lithofacies indicates that the facies was deposited in a marshy environment in a fluvial plane (Rashid et al 2013). The pollen results and the occurrence of pteridophytic and fungal spores support the hypothesis that the site was surrounded with a moist freshwater environment and that a forest was developed in the Madhupur tract. The ferric concentration of the walls of roots

Fig. 2.9 Stratigraphic description at Nayanipara

may reveal that the weathering acted upon the succession and that it experienced wet and dry weather.

2.3 Sea Level Changes from Tide Gauge Data

Tide gauge record refers to continuous record of height of tide, accurately observed at fixed interval of time and referred to a stable bench mark. First tide gauge was established in 1682 in Amsterdam. Only 122 stations have recorded from before 1900. Global sea-level change has been relatively well studied by making use of the coastal tide gauge data that are available (Woodworth and Player 2003) through the Permanent Service for Mean Sea Level (PSMSL). Tide-gauge records have been used extensively during this century to estimate sea level changes (Gornitz; Pugh 1987; Emery 1980; Barnet 1984; Aubrey and Emery 1993). At least a dozen attempts have so far been under taken during last 50 years to estimate the global sea level change based on tidal records. Tide-gauge records have been used extensively during this century to estimate sea level changes (Gornitz 1991; Emery 1980; Barnet 1984; Aubrey and Emery 1993). Studies of global mean sea level changes over the last 100 years based upon tide-gauge data yield values ranging between 0.5 and 3 mm/yr; most estimates lie within 1.0–2.0 mm/yr (Warrick and Oerlemans 1990).

References

Amieux P, Bernier P, Dalongeville R, Medwecki V (1989) Cathodoluminescence of carbonate-cemented Holocene beachrock from the Togo coastline (West Africa): an approach to early diagenesis. Sediment Geol 65:261–272

Aubrey DG, Emery KO (1993) Recent global sea level and land levels. In: Warrick RA, Barrow EM, Wigley TML (eds) Climate and sea level change: observation, projection and implication. Cambridge University Press, Britain, pp 45–56

Aucan J, Ridd PV (2000) Tidal asymmetry in creeks surrounded by salt flats and mangrove with small swamp slopes. Wetlands Ecol Manag 8:223–231

Barnett TP (1984) Recent changes in sea level and their possible causes. Clim Change 5:15–38

Beaman R, Larcombe P, Carter RM (1994) New evidence for the Holocene sea-level high from the inner shelf, central Great barrier Reef, Australia. J Sediment Res 64:881–885

Belperio AP (1979) Negative evidence for a Mid-Holocene high sea-level along the coastal plain of the Great Barrier Reef Province. Mar Geol 32:M1–M9

Birks HJB, Birks HH (1980) Quaternary palaeoecology. Edward Arnold Ltd, London, UK

Bryant EA, Young RW et al (1992) Evidence for Pleistocene and Holocene raised marine deposits, Sandon point, New-South-Wales. Aust J Earth Sci 39(4):481–493

Bunt JS, Williums WT, Bunt ED (1985) Mangrove species distribution in relation to tide at the seafront and up rivers. Aust J Mar Freshw Res 36:481–492

Chappell J (1993) Contrasting Holocene sedimentary geologies of lower Daly River, northern Australia, and lower Sepik-Ramu, Papua New Guinea. Sediment Geol 83:339–358

Clark RL, Guppy JC (1988) A transition from mangrove forest to freshwater wetland in the monsoon tropics of Australia. J Biogeogr 15:665–684

Cohen M, Filho P, Lara R, Behling H, Angulo R (2005) A model of Holocene mangrove development and relative sea-level changes on the Braganca Peninsula (Northern Brazil). Wetlands Ecol Manag 13:433–443

Dalrymple RW, Zaitlin BA, Boyd R (1992) Estuarine facies models: conceptual basis and stratigraphic implications. J Sediment Petrol 62:1030–1043

Devoy RJN (1979) Flandrian Sea-level changes and vegetational history of the lower Thames Estuary. Philos Trans R Soc London 285:355–407

Devoy RJN (1987) Sea-Level applications and management. Prog Oceanogr 18:273–286

Edwards Robin (2006) Sea levels: change and variability during warm intervals. Prog Phys Geogr 30(6):785–796

Ellison JC (1989) Pollen analysis of mangrove sediments as a Sea-level Indicator: assessment from Tongatapu, Tonga. Palaeogeogr Palaeoclimatol Palaeoecol 74:327–341

Emery KO (1980) Relative sea levels from tide-gauge records. Proc Natl Acad Sci USA 77(12):6768–6772

Encyclopædia Britannica Online (2012) Encyclopædia Britannica Inc., 2012. Web. 29 Jun. 2012. http://www.britannica.com/EBchecked/topic/532223/sedimentary-facies

Faegri K, Iversen J (1989) Text book of pollen analysis. Wiley, Chichester, UK

Fairbanks RG (1989) A 17000-year glacio-eustatic sea level record: influence of glacial melting rates on the Younger Dryas event and deep-ocean circulation. Nature 342:639–642

Fairbridge RW (1961) Eustatic changes in sea-level. Phys Chem Earth 4:99–185

Fujimoto K, Miyagi T (1993) Development process of tidal-flat type mangrove habitats and their zonation in the Pacific Ocean: a geomorphological study. Vegetatio 106:137–146

Geyh MA, Kudrass HR, Streif H (1979) Sea level changes during the Late Pleistocene and Holocene in the Strait of Malacca. Nature 278:441–443

Godwin H (1940) Studies of the post-glacial history of British vegetation, iii. Fenland pollen diagram, iv. Post glacial changes of relative land-and sea-level in the English Fenland. Philos Trans R Soc B230:239–303

Godwin H, Suggate RP, Willis EH (1958) Radiocarbon dating of the eustatic rise of ocean level. Nature 181:1518–1519

Gornitz V (1991) Global coastal hazards from future sea level rise. Palaeogeogr Palaeoclimatol Palaeoecol 89:379–398

Gornitz V (1993) Mean sea level changes in the recent past. In: Warrick RA, Barrow EM, Wigley TML (eds) Climate and sea level change: observation, projection and implication. Cambridge University Press, Britain, pp 25–44

Grindord J (1985) The palynology of mangroves on a prograde shore, Princess Charlotte Bay, North Queensland, Australia. J Biogeogr 12:323–348

Haq BU, Boersma A (1978) Introduction to Marine Micropaleontology. Elsevier, New York, p 376

Higgins CG (1994) Subsurface environment of beaches—temperature and salinity. In: Geologic Society of America, abstracts with programs, Seattle meeting, p A-364

Ireland S (1987) The Holocene sedimentary history of the coastal lagoon of Rio de Janiero state, Brazil. In: Tooley MJ, Shennan I (eds) Sea-level changes. Blackwell, Oxford

Jardine WG (1982) Determination of altitude. In: Van de Plassche OA (ed) Sea level research, manual for collection and evaluation of data. Geo Books, Norwich, 569–589

Jelgersma S (1961) Holocene sea-level changes in the Netherlands. Meded Geol Sticht Serie C 7:1–100

Kidson C (1981) Criteria for the evaluation of "evidence" of higher Holocene sea levels. Oceans 7:381–388

Laidler RB, Scott DB (1996) Foraminifera and Arcellacea from Porters Lake, Nova Scotia: modern distribution and paleodistribution. Can J Earth Sci 33:1410–1427

Larcombe P, Carter RM, Dye J, Gagen MK, Johnson DP (1995) New evidence for episodic post-glacial sea-level rise, central Great Barrier Reef, Australia. Mar Geol 127:1–44

Miall AD (1990) Principles of sedimentary basin analysis, 2nd edn. Springer, New York

Mildenhall DC (1994) Early to mid-Holocene pollen samples containing mangrove pollen from Sponge Bay, East Coast, North Island, New Zealand. J Roy Soc NZ 24:219–230

Palmer AJM, Abbott WH (1986) Diatoms as sea-level indicators. In: Van de Plassche O (ed) Sea-level research: A manual for the collection and evaluation of data. Geo Books, Norwich, pp 435–456

Preuss H (1979) Progress in computer evaluation of sea-level data within the IGCP project no. 61. In: Proceedings of the 1978 international symposium of coastal evolution in the Quaternary, Sao Paulo, Brazil

Pugh D (1987) Tides, surges and mean sea level. John Wiley & Sons, New York

Rashid T, Monsur MH, Suzuki S, OOI N (2009) Reconstruction of Holocene Paleo environment and evidence of Sea-level changes in the Bengal lowland. J Geogr 118(4):631–645

Rashid T, Suzuki S, Sato H, Monsur MH, Saha SK (2013) Relative sea-level changes during the Holocene in Bangladesh. J Asian Earth Sci 64:136–150

Retallack GJ (1997) A Colour Guide to Paleosols. John Wiley & Sons, Chichester (UK), 175 p

Scholl DW (1964) Recent sedimentary record in mangrove swamps and rise in sea level over the southwestern coast of Florida: Part 1. Mar Geol 1:344–366

Scholl DW, Stuiver M (1967) Recent submergence of southern Florida: a comparison with adjacent coasts and other eustatic data. Geol Soc Am Bull 78:437–454

Scoffin TP, Stoddart DR (1987) Beachrock and intertidal cements. In: Scoffin TP (ed) An Introduction to Carbonate Sediments and Rocks. Blackie Publishing Company, Glasgow, pp 401–425

Scott DB, Medioli FS (1980) Quantitative studies of marsh foraminiferal distribution in Nova Scotia: implications for sea level studies. Cushman Foundation for Foraminiferal Research (Special Publication No: 17)

Semeniuk V (1994) Predicting the effect of sea-level rise on mangroves in Northwestern Australia. J Coastal Res 10:1050–1076

Shennan I (1982) Interpretation of Flandrian sea-level data from the Fenland, England. Proc Geol Assoc 83(1):53–63

Shennan I (1986) Flandrian sea level changes in the Fenland. II. Tendencies of sea-level movement, altitudinal changes and local and regional factors. J Quat Sci 1:155–179

Shennan I, Tooley MJ, Davies MJ, Haggart BA (1983) Analysis and interpretation of Holocene sea-level data. Nature 302:404–406

Shennan I, Long A, Rutherford MM, et al. (1996) Tidal marsh stratigraphy, sea-level change and large earthquakes, a 5000 year record in Washington, U.S.A. Quatern Sci Rev 15:1023–1059

Shennan I, Long A, Rutherford MM, et al. (1998) Tidal marsh stratigraphy, sea-level change and large earthquakes, II, submergence events during the last 3500 years at Netarts Bay, Oregon, U.S.A. Quatern Sci Rev 17:365–393

Shepard FP (1961) Thirty-five thousand years of sea-level. In: Clement T (ed) Essays in marine geology in honour of K. O. Emery. Univ of South California Press, Los Angeles

Tooley MJ (1978) Sea-level changes during the last 9000 years in North-west England. Geog J 140:18–42

Tooley MJ (1982) Sea-level changes in North England. Proc Geol Assoc 93:43–51

Tooley MJ (1992) Recent sea-level changes. In: Allen JRL, Pye K (eds) Saltmarsh; morphodynamics, conservation and engineering significance. Cambridge University Press, Cambridge, pp 19–40

Turner, RJ (2005) Beachrock. In: Schwartz ML (ed) Encyclopedia of Coastal Science. Kluwer, The Netherlands, pp 183–186

Van de Plassche O (1986) Sea level research: a manual for the collection and the evaluation of data. Geobooks, Norwich

Van Harten D (1986) Use of ostracodes to recognise downslope contamination in paleobathymetry and a preliminary reappraisal of paleodepth of the Prasas Marls (Pliocene), Crete, Greece. Geology 14:856–859

Warrick RA, Barrow EM, Oerlemans J (1990) Sea level rise. In: Houghton JT, Jenkins GJ, Ephraums JJ (eds) Climate change: The IPCC scientific assessment. Cambridge University Press, Cambridge, pp 257–287

Woodroffe CD (1981) Mangrove swamp stratigraphy and Holocene transgression, Grand Cayman Island, West Indies. Mar Geol 41:271–294

Woodroffe CD, Thom BG, Chappell J (1985) Development of widespread mangrove swamps in mid-Holocene times in northern Australia. Nature 317:711–713

Woodroffe CD, Mulrennan ME, Chappell J (1993) Estuarine infill and coastal progradation, southern van Diemen Gulf, northern Australia. Sed Geol 83:257–275

Woodworth PL, Player R (2003) The permanent service for mean sea level: an update to the 21st century. J Coastal Res 19:287–295

Chapter 3
Global Sea-Level Changes Scenarios During the Holocene

3.1 Introduction

The sea level at any location changes constantly with changes in tides, atmospheric pressure, and wind conditions. Longer-term changes are influenced by changes in the Earth's climates. Consequently, the level is better defined as mean sea level, the height of the sea surface averaged over all stages of the tide over a long period of time (Britanica Concise Encyclopedia).

Sea level, the level of the sea, serves as the datum used for measurement of land elevations and ocean depths. Theoretically, one would expect sea level to be a fixed and permanent horizontal surface on the face of the earth, and as a starting approximation, this is true. However, a number of factors operate to cause variations in sea level ranging up to several meters from place to place and to cause long-term global variations, often severe enough to cause flooding and damage to coastal zones.

3.2 Different Thoughts and Scenarios of Global Sea Level Change During the Holocene

Fairbridge (1961) argued that the post-glacial rise of the sea was spasmodic and included regressive as well as transgressive phases. Additionally he argued that there were a number of stands of the sea in the late Holocene at heights up to 3.7 m above the present level at about 5700, 4900, 3700, 2400, 2200 and 1000 years BP.

Shepard (1963) indicated that there was general agreement on the nature of sea level change in the late glacial but that, in the later post glacial, this agreement broke down. He argued that in this later period the rise of sea level was a continuous one, at a rate diminishing in time but going on to the present day. He did not entirely rule out the possibility of a slightly higher than present late Holocene level but considered it to be unproven and drew attention to the fact that all

T. Rashid, *Holocene Sea-level Scenarios in Bangladesh*,
SpringerBriefs in Oceanography, DOI: 10.1007/978-981-4560-99-3_3,

Fairbridge's evidence for such higher levels came from Australia which remained, in his words, an "enigma".

There appeared a broad division into Fairbridge or Shepard schools of thought. Gill (1961), Mörner (1976), Hopley (1971), Gill and Hopley (1972), Fedorov (1972), Ters (1973) and Tooley (1974) all produced evidence from their respective regions supporting a sea level sequence as envisaged by Fairbridge. Conversely Jelgersma (1966), Bloom and Stuiver (1963), Bloom (1967), Hails and Hoyt (1971) and Kidson and Heyworth (1973) among others claimed to demonstrate the smooth exponential nature of sea level rise and denied that the late Holocene sea had ever exceeded that of the present.

Jelgersma (1966), in a review of sea level studies, discerned a third group favouring a steadily rising sea during the Holocene reaching its present level between 5000 and 3600 BP and subscribing to the standing sea level after 3600 BP theory (Fisk 1951; Godwin et al. 1958; Mc Farlan 1961). Guilcher (1969) drew attention both to the equivocal nature of much of the evidence and to the fact that even some of the leading disputants in the controversy did not regard the problems as solved.

Kidson argued that over the last 30 years the emphasis in studies of the recovery from low last glaciation sea levels has changed significantly. The search for a eustatic sea level curve having global relevance has ended. Studies into the rheology of the earth's crust, and recognition that the geoid has not remained stable over time, have resulted in the recognition that there must have been regional differences in eustatic response to deglaciation. As a part of this re-appraisal there has been a growing appreciation that crustal isostatic response to the removal of the weight of ice sheets has been accompanied by the later part of the post-war period attention has additionally been focused on the much greater potential for error over the whole field of paleoenvironmental reconstruction in the Holocene, including not only errors in dating, but also a large number of scientists are withholding judgement on the nature of sea level rise and, more particularly, on the problem of higher than present late Holocene eustatic sea levels. The problems of uncertainty in the early 1960s have not yet been resolved but the bases of uncertainty have changed.

With regards to Mid-Holocene High Stand (MHHS), Woodroffe and Horton (2005) showed that despite being a large geographical region, the nature of Holocene sea-level change is broadly similar in all locations. Differences do exist in the timing and magnitude of the MHHS and the nature of late Holocene sea level fall across the region. They argued that when the Indo-Pacific is divided into smaller regions, these discrepancies do not disappear, and in some cases the discrepancies are large within a single coastline. They showed sea-level reconstruction records of the countries belong to Indo-Pacific region, particularly high stand in terms of timing and magnitude (Table 3.1).

Ingolfsson and Hjort (1999) found distinct differences between the deglaciation histories of the large Northern Hemisphere ice sheets and Antarctica. Between 10 and 8 Kya the meltwater input from the Northern Hemisphere ice sheets diminished rapidly, as they disintegrated and melted away. There were minimal or no

Table 3.1 Global sea level records during Holocene period

Indo-Pacific Region	Location and authors	Shoreline indicators used and indicative range(if known)	Tectonic activity/tidal range etc.	Dating method	Timing of mid-Holocene high stand	Magnitude of mid-Holocene high stand
SW Indian Ocean	South Africa (Ramsay and Cooper 2002)	Beachrock (± 0.2 m), barnacles and attached oysters (± 0.5 m), estuarine fill sequences	Stable	Uranium series (beachrock)	1HS-4650 C14 yrs BP; 2HS-1610 C14 yrs BP	1HS-3.5 m above SL; 2HS-1.5 m above SL
SW Indian Ocean	S Africa (Compton 2001)	Saltmarsh organic materials/ mollusc	Stable, microtidal range	Radiocarbon	1HS-6800 C14 yrs BP; 2HS-1300 C14 yrs BP	1HS-1 to 3 above SL; 2HS-0.5 to 1 m above SL
SW Indian Ocean	Mauritius, Reunion and Mayotte (Colonna et al. 1996; Camoin et al. 1997)	Fringing coral reefs	Stable	Uranium series	SL stabilized at current levels at 3 ka cal BP	No HS record
Northern Indian Ocean	Sri Lanka (katupotha and Fujiwara 1988)	Coral and marine shells	Stable	Radicarbon	1 HS-6170 to 5100 C14 yrs BP; 2HS-3210 to 2330 C14 yrs BP	1HS- 1 m above SL; 2HS-1 m (may not have lowered btwn 2 heights)
Northern Indian Ocean	East coast of India (Banerjee 2000)	Beach ridges (± 0.5 m), porites coral colony, intertidal shells	Stable	Radiocarbon (shells) Uranium series)	1HS-7300 to 5660; 2HS-4300 to 2500	1HS-5 to 3 m; 2HS-3 m above SL
Northern Indian Ocean	Bangladesh (Islam and Tooley 1999)	Sedimentary sequences and microfossils	Possible tectonic activity/ glacio isostacy	Radiocarbon	No evidence for SL HS	No evidence
SE Asia	Malay–Thai peninsula (Tija 1996)	Malaysia-Abrasion platforms, sea-level notches, oyster beds; Thailand-peat, marine shells, mangrove woods	Stable assumes TL not changed	Radiocarbon	Malay-HS-5000 and 2800 c14 yrs BP; Thai-6000, 4000, 2700 c14 yrs BP	Malay-5 and 3 m; Thai-4, 2.5 and 2 m above SL

(continued)

Table 3.1 (continued)

Indo-Pacific Region	Location and authors	Shoreline indicators used and indicative range(if known)	Tectonic activity/tidal range etc.	Dating method	Timing of mid-Holocene high stand	Magnitude of mid-Holocene high stand
SE Asia	Sunda Shelf, Indonesia (Handebuth et al. 2000)	Sediment facies from cores	Stable	Radiocarbon	Record from 12,000 to 14000 cal yrs BP-shows MWP 1a (16 m SL rise in 300 yrs)	N/A
SE Asia	Phuket, S Thailand (Schoffin and Le Tissier 1998)	Coral microatolls	Stable	Radiocarbon	Before 6000 cal yrs BP	1 m above SL
SE Asia	Huon Peninsula, P. New Guinea (Chappell and Polach 1991)	Raised coral terraces	Tectonically active (emergence btwn 0.5 and 3 m/ 1000 yrs)	Uranium series	Late glacial/early Holocene record of SL rise up to 5800	N/A
SE Asia	Cocos keeling Islands (Woodroffe and McLean 1990)	Coral microatolls	Stable	Radiocarbon	Since about 3000 c 14 yrs BP	At least 0.5 m above SL
SE Asia	Strait of Malacca (Geyh and Kudrass 1979)	Mangrove wood and peat (indicative meaning not accounted for)	Stable	Radiocarbon	Approx. 4980 c14 yrs BP	2.5–5.8 m above SL
Northern Asia	Japan (Nakada et al. 1991)	Intertidal and subtidal shells, wood and peat	Tectonically active	Radiocarbon	HS at 6000	5 m above SL
Northern Asia	Japan (Sawai et al. 2002)	Sedimentary sequences and microfossils	Tectonically active	Radiocarbon	3 falls in RSL since 3000 cal yrs BP	N/A
Northern Asia	Southeast China (Yim and Huang 2002)	Fixed biological indicators (oyster beds)	Local monsoon activity	Radiocarbon	5140 c14 yrs BP	No more than 2 m

(continued)

Table 3.1 (continued)

Indo-Pacific Region	Location and authors	Shoreline indicators used and indicative range(if known)	Tectonic activity/tidal range etc.	Dating method	Timing of mid-Holocene high stand	Magnitude of mid-Holocene high stand
Australia	Central Grest Barrier Reef, Australia (Beaman et al. 1994; Larcombe and Carter 1998)	Oyester beds	Stable, some differential flexing of shelf over past 6000 yrs	Radiocarbon	Btwn 5660 and 4040 c14 yrs BP	1.65 m above SL
Australia	Eastern Australia (Flood and Frankel 1998)	Intertidal tube worms	Stable	Radiocarbon	Sometime before and after 3420 c14 yrs BP	At least 1 m
Australia	Bonaparte Gulf, NE Australia (Yokohama 2000)	Sedimentary facies and microfossils	Stable	Radiocarbon	LGM and early late glacial	N/A
Australia	New south Wales (Baker et al. 2001)	Fixed biological indicators (tubeworms, barnacles, oyesters0	Stable	Radiocarbon	Around 3900 cal yrs BP	At least 1 m above SL
Pacific	Fiji (Nunn and Peltier 2001)	Coral (Porites) microatolls and intertidal shells from raised beaches	Hotspot activity over Holocene timescales (subsidence)	Radiocarbon	1HS-5650 to 3200 c14 yrs BP; 2HS-6100 to 4550 and 3590–2800 c14 yrs BP	1HS-1.35 to 1.5 m above SL; 2HS-(i) 0.75–1.85 m, (ii) 0.90–2.46 m above SL
Pacific	Taihiti (Bard et al. 1996)	Coral (Acropora)	Slight subsidence (−0.2 mm/yr)	Uranium series	Record from 14000 to 5000 cal yrs BP	N/A
Pacific	W Central Equitorial Pacific (Grossman et al. 1998)	Coral, microatolls, beachrock, mollusks, peat	Hotspot activity over Holocene	Radiocarbon	4 ka c14 yrs BP	1–2 m above SL

meltwater inputs from the Cordilleran Ice Sheet and the Barents Sea-Kara Sea ice sheets after 10–9.5 Kya. The Scandinavian Ice Sheet continued contributing meltwater until ca. 8.5 Kya. Melt water input from the Greenland Ice Sheet was minimal after 8 Kya. The Laurentide Ice Sheet contributed large volumes of meltwater until ca. 8 Kya, but its input dwindled rapidly between 8 and 7 Kya and was minimal after 7 Kya. The global glacio-eustatic sea level rise, however, continued until some between 6 and 5 Kya. The continued deglaciation of currently ice-free land in Antarctica occurred gradually between ca. 8 and 5 Kya. A large southern portion of the marine-based Ross Ice Sheet disintegrated during this late deglaciation phase. Some currently ice-free areas were deglaciated as late as 3 Kya. As the terrestrial field evidences from Antarctica suggests that deglaciation there was distinctly delayed compared to the Northern Hemisphere deglaciation, they suggest that this late Antarctic deglaciation caused part of the global sea level rise between 8 and 7 Kya and most of it between 7 and 5 Kya. They concluded that the global mid-Holocene sea level high stand, broadly dated between 8 and 4 Kya, and the Littorian-Taped transgressions in Scandinavia and simultaneous transgressions recorded from sites e.g. in Svalbard and Greenland, dated to 7–5 Kya, probably reflect input of meltwater from the Antarctic deglaciation.

Isla (1989) argued that if rising sea levels dominate in the northern hemisphere (NH), falling or fluctuating sea levels predominate in the southern hemisphere (SH). Endogenic processes (tectonics, isostasy or geoidal changes) could explain local or regional mean sea level (MSL) fluctuations but not a hemispherical one. Evidence from South America, Africa, Antarctica, Australia and the Pacific and Indian Oceans suggest that the Holocene transgression rose above the present MSL, in higher latitudes before the tropics. By plotting latitude against the age of MSL arrival at present coasts, good correlation is observed. Oceanic salinity mixing has been already proposed to explain this mid-Holocene sea-level fluctuation. Climate could be the only factor responsible for this hemisphere-wide behavior of MSL. It has been suggested previously that the climate of the SH precedes that of the NH by 3000 years. The climatic optimum, or maximum warmth, had been occurred predominantly about 6000 years BP in the NH, but about 10–9000 years BP in the SH. He found that short-term climatic effects on the sea level (monsoons, southern oscillation/El Niño phenomena) should have significant occurrences during the past in the windiest oceanic hemisphere. This latitudinal trend in former MSL should be considered when using shorelines as reference points for measuring vertical crustal movements. The author found similar sea level history of Brazilian coast and Senegal coast of Africa. The sea-level high stand records are summarized here (Table 3.2).

The author shows a relationship between latitude and age of maximum transgression which shows the oldest evidence for Holocene MSL in the southern hemisphere and shows younger ages towards the equator. He concluded with four following findings:

Table 3.2 Global sea level high stand records

Region	Location and author	Shoreline indicators (if known) and dating method	Timing of mid-Holocene high stand	Magnitude of mid-Holocene high stand	Regressive phase (if known)
South America: Brazil	Santa Catarina (28°S) to Salvador de Bahia (13°S) (Martin et al. 1985)	N/A	HS- 5150 BP	5 m at Salvador; 2.3 m at Paranagua; 3.5 m at Itajai-Laguna	4000 and 2700 BP
Brazil	Chile (Paskoff 1978, 1980)	Beach shells, radiocarbon	HS-3700 ± 120 (at Herradura Bay) and 4400 ± 120 BP (Los Molles)	4–5 m higher at along the coast of Chile	
Brazil	Buenos Aires Province (35°S) (Schnack et al. 1982)		5000 years BP	2.2–2.5 m	
Brazil	Rio Quequen inlet (38°34′S) (Isla et al. 1986)		7640 +/= 90 BP	2.5 m above SL	
	Bahia Blanca estuary (39°S) (Gonzalez et al. 1983)		Some time before 6000 BP		
	Colorado delta (40°S)	Beach ridges	6000 BP		
	Pantagonian coast (Codignotto 1983)	Ac-143 Gravel-Ac-160	3310 ± 90 BP (at Puerto Lobos −42°S); 5720 ± 105 BP (at Valdes- spit 42°S)	12 m	
	Puerto del Hambre (Magallanes Strait, 53°30′ S) (Porter et al. 1984)	Marine clay with foraminifera and dinoflagellates	Between 7980 ± 50 and 3970 ± 70 BP (QL-1467)	2–3.5 m	

(continued)

Table 3.2 (continued)

Region	Location and author	Shoreline indicators (if known) and dating method	Timing of mid-Holocene high stand	Magnitude of mid-Holocene high stand	Regressive phase (if known)
	Beagle Channel-Punta Pinguinos (Auer 1959)	Peat	7450 ± 100 BP (Y-191) and 7660 ± 100 BP (Y-190)	5.4 m	
	Argentine border at Bahia Lapataia (54°50′ S) (Rabassa et al. 1986)	*Mytilus* shells	8240 ± 60 BP (marine transgression)	Altitude of 1.65 m above MSL	
	Rio Ovando and Isla El Salmon (Rabassa et al. 1986)		4425 and 3860 BP	8 m above SL	
Africa	Coast of Mauritania (Nouakchottian transgression, 18°N; Faure and Elouard 1967)	Stromatolithic algae	5500 BP	3 m above SL	Low SL of – 3.5 m at approx. 4100 BP
	Senegal coast (Tanma Lake 14°55′N)		5650 ± 130 BP 2630 BP (Lezine et al. 1985)	0.5 m	Low level at 4000 BP
	Gulf of Guinea (4°30′–6°30′N) (Tastet 1978)	Elevated and fossil peaty mangrove, radiocarbon	(i) Older than 5700 BP and younger than 4900 BP; (ii) btwn 3400 and 1400 BP (iii) 760–270 BP		4000 BP Btwn 1120 and 970 BP (Benia)
	Sao Thome and Annobon (equator) (Cornen et al. 1977)	Elevated beaches	3900 BP; 2300 BP; 1430 BP	MSL 0.5 m above present; ±1.5 m; Above MSL	
	Koilou estuary-Congo (Moguedet et al. 1986; Cornen et al. 1977)		Marine transgression at 9000 BP; 5000 BP	SL 12 m below present;	

(continued)

Table 3.2 (continued)

Region	Location and author	Shoreline indicators (if known) and dating method	Timing of mid-Holocene high stand	Magnitude of mid-Holocene high stand	Regressive phase (if known)
	Praia das Conchas-Mossamedes Bay, Angola (Cornen et al. 1977; Giresse et al. 1984)	Rocky littoral fauna	3430 ± 100 BP	Close to present MSL 5 m above SL	3040 ± bp at 3 m above SL
	Groenvlei (S Africa-34°S) (Butzer and Helgren 1972)		6870 ± 160	2.5 m above SL	Drop to present SL before 1905 ± 60 BP
Oceania, Australia and Indian Ocean	Guam-Marianas Islands, 13°N (Curray et al. 1970)	Coral	2880 BP at Toguan Bay	0.6 m	
	Southern Marianas (Ida et al. 1984)	*Tridacna* shells and corals; radiocarbon	Btwn 5115 ± 100 to 2900 BP	Maximum hight of the samples was 0.6 m above MSL although there are emerged reefs up to +1.35 m (Agfayan Bay and benches and notches at heights of 2 and 4.5 m	
	Rota Island (Ida et al. 1984)	*Porites*- Coral	4510 ± 140 BP	2.82 m above MSL	
	Adorius Island (West carollines, 7°N) Eston and Ku 1980)	*Corals*	6500 ± 200 BP and 5133 ± 80 BP	2 m above SL	

(continued)

Table 3.2 (continued)

Region	Location and author	Shoreline indicators (if known) and dating method	Timing of mid-Holocene high stand	Magnitude of mid-Holocene high stand	Regressive phase (if known)
	Strait of Malacca – 2°30′N to 1°30′N)	Shells, corals beachrock, mangrove woods and peat	Btwn 4400 ± 65 BP and 4040 ± 100 BP	5.90 m	Sl drop of – 2 m at 2000 BP
	Laccadive Archipelago – 8 to 12°N (Siddiquie 1980)		?	3 m above MSL	
	Southwestern and southern Sri Lanka – 6 to 7°N (Katupotha and Fujiwara 1987)	Shells and corals	Btwn 6200 and 5100: btwn 3200 and 2300 BP	1 m higher	4000 BP, SL fell below present
Australia	St. Helena Island (Lovell 1975)	Shells, radiocarbon	2540 ± 85 BP	2 m	
	Bowen and Palm Islands, Queensland −19 to 20°S (Hopley 1984) (Chappell 1983)	Mangroves, corals and shells radiocarbon	MSL rise of 10 mm/yr btwn 9000 and 8000 BP and stability for the last 6500 BP;		1 m drop in the last 6000
New Zealand	New Zealand −34 to 46°S (Gibb 1983)		6500 BP;		
	Wainui beach		7590 ± 90 BP	3.6 m above MSL	
	Pakarae River		6740 ± 150 BP	22.5 m	
	Hicks Bay		6710 ± 80 BP	8.3 m	
	New Hebrides Islands (Bloom 1980)		7100–5700 BP	9 and 13 m at Santos(15°30′S); 3 and 6 m at Efate (18°S)	

(continued)

Table 3.2 (continued)

Region	Location and author	Shoreline indicators (if known) and dating method	Timing of mid-Holocene high stand	Magnitude of mid-Holocene high stand	Regressive phase (if known)
	Araki Island (Bloom and Yonekura 1985);		5500 BP	26 m above (4.75 mm/yr uplift);	
	Tangoa Island and South Santo Island;		4500 BP	19 m (5.4 mm/yr);	
	Tasmaloum Peninsula (santo)		6000 BP	33 m (7.9 mm/yr)	
	New Caledonia −22°S (Launay and Recy 1972)	Coral	7360 ± 170 BP	1.14 m	
	Marais de Mara	peat	5600 ± 150 BP	0.14 m	
	Viti Leuvu (Fiji Islands −18°S) (Sugimura et al. 1982; Matsushina et al. 1984)	Shells- *Chama pacifica*	3000–2000 BP	1.6 m	4400 ± 90 BP −1.2 m below MSL
	Tonga Island −21°S ?		Ca. 6200 BP	2.2 m	
	Cook Islands-18 to 22°s (Yonekura et al. 1983)	Corals	3410 ± 170 BP; 5020 ± 190 BP	1.8 m; 1.2 m above SL	
	Aitutaki −18°50′S and Roratonga −21°S (Schofield 1970)	Corals	6000 BP (at Rarotonga); 4700 BP on Aitutaki	Higher than −1.4 m; higher than −0.7 m	
	Society Islands −16 to 17°S (Pirazzoli 1985)	Corals	Older than 3400 BP	0.9 m	

(continued)

Table 3.2 (continued)

Region	Location and author	Shoreline indicators (if known) and dating method	Timing of mid-Holocene high stand	Magnitude of mid-Holocene high stand	Regressive phase (if known)
	Tuamotu Islands (French Polynezia-15°S) (Pirazzoli and Montaggioni 1986)	Conglomerates, corals and *Tridacna* shells	5000 BP	0.9 m	Falling about 1200 BP
Antarctica	Shetland del Sur Islands 62°S (Sugden and John 1973)	Raised beach	9670 ± 230 BP		
	James Ross Island –64°S (Rabassa 1982)	Broken and articulated shells	5175–5915 BP	35–40 m	
	East Ongul Island –69°S (Meguro et al. 1964)	Fossil mollusca (*Adamussium colbecki*)	3840 ± 110 BP	3–4 m above SL	

(1) During the Holocene, MSL was higher than present in the southern hemisphere.
(2) If postglacial deformation is the dominant effect that controlled MSL trends in the NH, climatically-triggered phenomena ("true"glacial-eustacy, oceanic or meteorological effects) were responsible for the MSL records of the SH.
(3) The latitudinal lags recognized in the MSL records should be further defined by characteristics of climate variations.
(4) Such delays in eustatic trends must be evaluated when using MSL records for tectonic calculations.

References

Auer V (1959) The Pleistocene of Fuego-Patagonia: Part III: Shorelines Displacements. Annales Academiae Scientiarum Fennicae Series A 3. Geologiska-Geographica, Helsinki, p 247

Baker RGV, Haworth RJ, Flood PG (2001). Inter-tidal fixed indicators of former Holocene sea-levels in Australia: a summary of sites and a review of methods and models. Quatern Int 83–85:257–273

Banerjee PK (2000) Holocene and Late Pleistocene relative sea level fluctuations along the east coast of India. Mar Geol 167:243–260

Bard E, Hamelin B, Arnold M, Montaggioni L, Cabioch G, Faure G, Rougerie F (1996). Deglacial sea-level record from Tahiti corals and the timing of global meltwater discharge. Nat 382:241–244

Beaman R, Larcombe P, Carter RM (1994) New evidence for the Holocene sea-level high from the inner shelf, central Great barrier Reef, Australia. J Sediment Res 64:881–885

Bloom AL (1967) Pleistocene shorelines: a new test of isostacy. Geol Soc Am Bull 78:1477–1493

Bloom AL (1980) Late quaternary sea level change on south pacific coasts: a study in tectonic diversity. In: Nils-Axel Morner (ed) Earth Rheology, Isostasy, and Eustasy. Wiley-Interscience, New York, pp 505–516

Bloom AL, Stuiver M (1963) Submergence of the Connecticut coast. Science 139:332–334

Bloom AL, Yonekura N (1990) Graphic analysis of dislocated quaternary shorelines. National Academic Press, Washington, DC, pp 104–115

Butzer KW, Helgren DM (1972) Late cenozoic evolution of the cape coast between Knysna and Cape St. Francis, South Africa. Quatern Res 2(2):143–169

Camoin GF, Colonna M, Montaggioni LF, Casanova J, Faure G, Thomassin BA (1997) Holocene sea level changes and reef development in the southwestern Indian Ocean. Coral Reefs 16:247–259

Chappell J, Polach H (1991) Postglacial sea-level rise from a coral record at Huon Peninsula, Papua New Guinea. Nat 349:147–149

Chappell J, Chivas A, Wallensky E, Polach HA, Aharon P (1983) Holocene palaeo-environmental changes, central to north Great Barrier Reef inner zone. BMR J Aust Geol Geophys 8:223–235

Chappell J (1993) Contrasting Holocene sedimentary geologies of lower Daly River, northern Australia, and lower Sepik-Ramu, Papua New Guinea. Sediment Geol 83:339–358

Codignotto JO (1983) Depósitos elevados y/o acreción Pleistoceno-Holoceno en la costa fueguino patagónia. Simp. Oscilaciones del Nivel del Mar el Ultima Hemiciclo Deglacial en la Argentina. CIC, Actas: 12–26, Mar del Plata

Colonna M, Casanova J, Dullo WC, Camoin G (1996) Sea-level changes and d180 record for the past 34,000 years from Mayotte Reef, Indian Ocean. Quatern Res 46:335–339

Compton JS (2001) Holocene sea-level fluctuations inferred from the evolution of depositional environments of the southern Langebaan Lagoon salt marsh, South Africa. Holocene 11:395–405

Curray JR, Francis PS, Herbert HV (1970) Late Quaternary sea-level studies in Micronesia: carmarsel expedition. Geol Soc Am Bull 81:1865–1880

Fairbridge RW (1961) Eustatic changes in sea-level. Phys Chem Earth 4:99–185

Fedorov PV (1972) Postglacial transgression of the black sea. Int Geol Rev 14(2):160–164

Fisk HN (1951) Loess and Quaternary geology of the lower Mississipi Valley. Jour. Geol. 59:333–356

Flood PG, Frankel E (1989) Late Holocene higher sea level indicators from eastern Australia. Mar Geol 90:193–195

Geyh MA, Kudrass HR, Streif H (1979) Sea level changes during the Late Pleistocene and Holocene in the Strait of Malacca. Nature 278:441–443

Gibb JG (1983) Sea levels during the past 10000 years B.P. from New Zealand region-South Pacific Ocean. In: International symposium of coastal evolution during the Holocene, Abstracts, Komazawa University, Tokyo, pp 28–31

Gill ED (1961) Changes in the level of the sea relative to the land in Australia during the Quaternary era. Z Geomorphol Suppl 3:73–79

Gill ED, Hopley D (1972) Holocene sea levels in eastern Australia, a discussion. Mar Geol 12:223–242

Giresse P, Hoang CT, Kouyoumontzakis G (1984) Analysis of vertical movements deduced from a geochronological study of marine Pleistocene deposits, southern coast of Angola. J Afr Earth Sci 2(2):177–187

Godwin H, Suggate RP, Willis EH (1958) Radiocarbon dating of the eustatic rise of ocean level. Nature 181:1518–1519

González M, Panarello H, Marino H, Valencia S (1983) Niveles marinos del holoceno en el estuario de bahía blanca (Argentina). Isótopos estables y microfósiles calcáreos como indicadores paleoambientales, Proceedings Simposio Oscilaciones del Nivel del Mar Durante el Ultimo Hemiciclo Deglacial en Argentina. Mar del Plata, pp 48–68

Grossman EE, Fletcher III CH, Richmond BM (1998) The Holocene sea-level high stand in the equatorial Pacific: analysis of the insular paleosea-level database. Coral Reefs 17:309–327

Guilcher A, Leopold B, Francois D et al (1969) Les recif et lagons corlliens de Mopelia et de Bora-Bora et quelques autres recifs et lagons de Comparaison, (Tahiti, Scilly, Tuamotu occidentals): morphologic, sedimentologie, fonctionnement hydrologique. Memoires OSTROM, No. 38, Paris, p 103

Hails JR, Hoyt JH (1971) The question of Late Quaternary changes of sea level in New South Walse, Australia. Quatwenaria 14:255–264

Hanebuth TJJ, Stattegger K, Grootes PM (2000) Rapid flooding of the Sunda Shelf: a late-glacial sea-level record. Science 288:1033–1035

Hopley D (1971) Sea-level and environmental changes in the Late Pleistocene and Holocene in North Queensland, Australia. Quaternaria 14:159–166

Hopley D (1982) Geomorphology of the Great Barrier Reef: quaternary development of Coral Reefs, John Wiley-Interscience: New York, p 453

Ingolfsson Olafur, Hjort Christian (1999) The Antarctic contribution to Holocene global sea level. Polar Res 18:323–330

Isla FI (1989) Holocene sea-level fluctuation in the Southern Hemisphere. Quat Sci Rev 8:359–368

Isla FI, Ferrero L, Fasano JL, Espinosa MA, Schnack EJ (1986) Late Quaternary marine estuarine sequences of the Southeastern coast of the Buenos Aires Province, Argentina. Quat S Am Antarct Peninsula 4:137–157

Islam MS, Tooley MJ (1999) Coastal and sea-level changes during the Holocene in Bangladesh. Quatern Int 55:61–75

Katupotha J, Fujiwara K (1988) Holocene sea level change on the Southwest and South Coast of Sri Lanka. Palaeogeogr Palaeoclimatol Palaeoecol 68:189–203

Kidson, Heyworth A (1973) The Flandrian sea-level rise in the Bristol Channel. Proc Ussher Soc 2:565–584

Larcombe P, Carter RM (1998) Sequence architecture during the Holocene transgression: an example from the Great Barrier Reef shelf, Australia. Sediment Geol 117:97–121

Launay J, et Recy J (1972) Variations relatives du niveau de la mer et néotectonique en Nouvelle-Calédonie au Pléistocène supérieur et à l'Holocène. Rev Géogr Phys Géol Dyn 14(1):47–65

Lézine A-M, Bieda S, Faure H, Saos JL (1985) Etude palynologique et sédimentologique d'un milieu margino-littoral: la tourbière de Thiaye (Sénégal). Sci géol Bull 38(1):79–89

Lovell ER (1975) Evidence for a higher sea level in Moreton Bay, Queensland. Mar Geol 18:M87–M94

Martin L, Flexor JM, Blitzkow D, Suguio K (1985) Geoid change indications along the Brazilian coast during the last 7000 years. In: Proceedings of the fifth International Coral Reef Congress, Tahiti vol 3, pp 85–90

Matsushima Y (1984) Shallow marine molluscan assemblages of Postglacial period in the Japanese Islands-its historical and geographical changes induced by the environmental changes. Bull Kanagawa Pref Mus (Nat Sci) 15:37–109

Mc Farlin E (1961) Radiocarbon dating of late Quaternary deposits, South Louisiana. Geol Soc Am Bull 72:129–158

Meguro H, Yoshida Y, Uchio T, Kigoshi K, Sugawara K (1964) Quaternary marine sediments and their geological dates with reference to the geomorphology of Kronprince Olav Kyst (in Antarctic Geology., ibid.), pp 73–80

Mörner N-A (1976) Eustasy and geoid changes. J Geol 84:123–151

Nakada M, Yonekura N, Lambeck K (1991) Late Pleistocene and Holocene sea-level changes in Japan: implications for tectonic histories and mantle rheology. Palaeogeogr Palaeoclimatol Palaeoecol 85:107–122

Nunn PD, Peltier WR (2001) Far-field test of the ICE-4G model of global isostatic response to deglaciation using empirical and theoretical Holocene sea-level reconstructions for the Fiji Islands, Southwestern Pacific. Quatern Res 55:203–214

Paskoff R (1978) Sur l'evolution geomorphologique du grand escarpement cotier du desert chilien. Geogrophie Physiqeel Quaternarie 4:351–360

Porter SC, Stuiver M, Heusser CJ (1984) Holocene sea-level changes along the strait of Magellan and Beagle channel, southernmost South America. Quatern Res 22:59–67

Rabassa J, Heusser C, y Stuckenrath R (1986) New data on Holocene sea transgression in the Beagle Channel: Tierra del Fuego, Argentina. In: En Rabassa, J. (ed) Balkema, Quaternary of South America and Antarctic Peninsula, Rotterdam, 4:291–309

Sawai Y, Nasu H, Yasuda Y (2002) Fluctuations in relative sea-level during the past 3000 yr in the Onnetoh estuary, Hokkaido, northern Japan. J Quatern Sci 17:607–622

Scoffin TP, Le Tissier MDA (1998) Late Holocene sea level and reef-flat progradation, Phuket, South Thailand. Coral Reefs 17:273–276

Shepard FP (1963) Thirty-five thousand years of sea-level (in Essays in marine geology, in honor of K.O. Emery, ed. by Clements, T.). Los. Univ. South. Cal. Press pp 1–10

Schofield JC (1970) Notes on Late Quaternary Sea levels, Fiji and Ranatonga. NZ J Geol Geophys 13(1):199–206

Siddiquie HN (1980) The ages of the storm beaches of the Lakshadweep (Laccadives). Mar Geol 38:M11–M20

Sugden DE, John BS (1973) The ages of glacier fluctuations in the South Shetland Islands, Antarctica. In: E.M. van Zinderen Bakker (ed) Palaeoecology of Africa, the surrounding Islands, and Antarctica vol 8, Balkema, Cape Town, pp 139–159

Tastet JP (1981) Morphogie des littoraux sedimentaries liee aux variations du niveau de la mer: example du Golfe de Guinee. Ocean 7(4):455–472

Ters M (1973) Les variantions du niveau marin depuis 10000 ans, le long du littoral atlantique frncais. In: Le Quatrnaire: geodynamique, stratigraphie et environment. Suppl Bull Assoc Fr Et Quat 36:114–135

Tija HD (1996) Sea-level changes in the tectonically stable Malay-Thai Peninsula. Quatern Intern 31:95–101

Woodroffe CD, McLean R (1990) Microatolls and recent sea-level change on coral atolls. Nat 344:531–534

Woodroffe SA, Horton BP (2005) Holocene sea-level changes in the Indo-Pacific. J Asian Earth Sci 25:29–43

Yim WW-S, Huang G (2002) Middle Holocene higher sea-level indicators from the South China coast. Mar Geol 182(3–4):225–230

Yokoyama Y, Lambeck K, De Deckker P, Johnston P, Fifield LK (2000) Timing of the Last Glacial maximum from observed sea-level minima. Nat 406:713–716

Yonekura N, Ishii T, Saito Y, Maeda Y, Matsuhima Y, Matsumoto E, Kayanne H (1988) Holocene fringing reefs and sea-level change in Mangaia Island, Southern Cook Islands. Palaeogeogr Palaeoclimatol Palaeoecol 68:177–188

Chapter 4
Landforms Development in Bangladesh

4.1 Historical Evolution of Bangladesh Since Mesozoic Era

Bangladesh is divided into two major tectonic units: (i) Stable Pre-Cambrian Platform in the northwest and (ii) Geosynclinal basin in the southeast. A third unit, a narrow northeast-southwest trending zone called the hinge zone separates the above two units almost through the middle of the country. This hinge zone is currently known as palaeo continental slope (Banglapedia 2003) (Fig. 4.1).

Bengal Geosyncline is one of the world's largest geosyncline that includes the Bengal Basin and the Bay of Bengal. The evolution of the Bengal Basin started in the Permo-Carboniferous with the sedimentation in the faulted Gondwana Basins. The break-up of the Gondwanaland in the Cretaceous Period and the marine transgression led to the sedimentation in the Bengal Geosyncline.

Bengal Basin, occupied mostly by Bangladesh, is the result of plate collision between the Indian plate and the Asian plate.

Before Cretaceous time (125 mya), the Indian plate including part of Bangladesh (greater Rangpur and Dinajpur area) was joined together with Antarctica, Africa, Australia and South America, forming a super continent called Gondwanaland.

The remaining part of Bangladesh did not exist at that time. It was only after the rifting, northward drifting of the Indian plate and the ultimate collision with the Asian plate that the Himalayas were formed and the deltaic plain of Bangladesh created.

During the early Tertiary period (Mid-Palaeocene-59 mya) "soft collision" started between India and Southeast Asia. Most of Bangladesh was under open marine condition. The Tethys Sea was disappeared.

During the late Tertiary period (44 mya) (Middle Eocene) "hard collision" started between India and South Asia. The open sea retreated to the south in response to the uplift of the Himalayas in the north and river systems built out a large deltaic land that formed the backbone of the present Bengal delta.

Since the Oligocene time (35 mya), plate collision continued and major SEDIMENTS were shed by the rising Himalayas as large river systems started filling up the proto-Bengal basin to the south.

T. Rashid, *Holocene Sea-level Scenarios in Bangladesh*,
SpringerBriefs in Oceanography, DOI: 10.1007/978-981-4560-99-3_4,

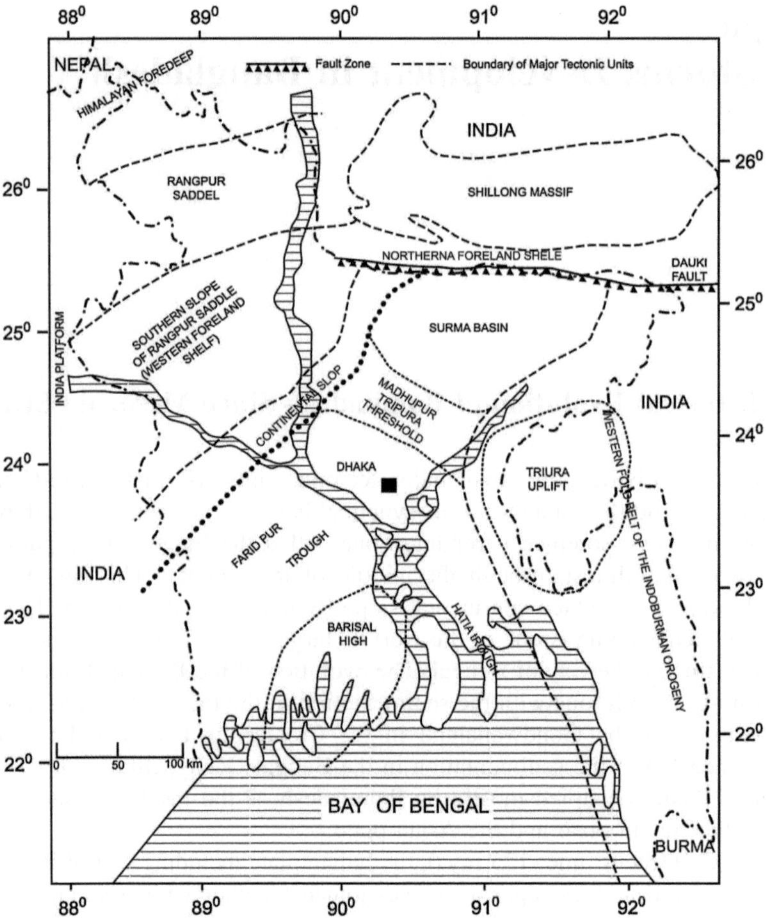

Fig. 4.1 Tectonic structure of Bangladesh (partly after GUHA, 1978)

From Miocene onward (from 25 mya), rapid rise in the Himalayas accompanied by rapid SUBSIDENCE in the basin resulted huge in deposition of a huge sedimentary pile with the simultaneous development of a mega-delta. This delta building continues till now as Ganges–Brahmaputra delta.

About 22 Ma, Early Miocene, a major collision occurred between India and South Tibet in the north and India and Burma in the east.

In response to the eastwards directed subduction of the Indian Plate, the molasses sediment (The term "molasse" refers to the sandstones, shales and conglomerates formed as terrestrial or shallow marine deposits in front of rising mountain chains) of this zone were folded into a series of elongated asymmetric folds arranged in en-echelon form. The alignment of the folds follows NNW-SSE in the Chittagong hill tracts. During this time the direction of the Indian plate convergence changed from north to north–east with increasing collision with Southeast Asia.

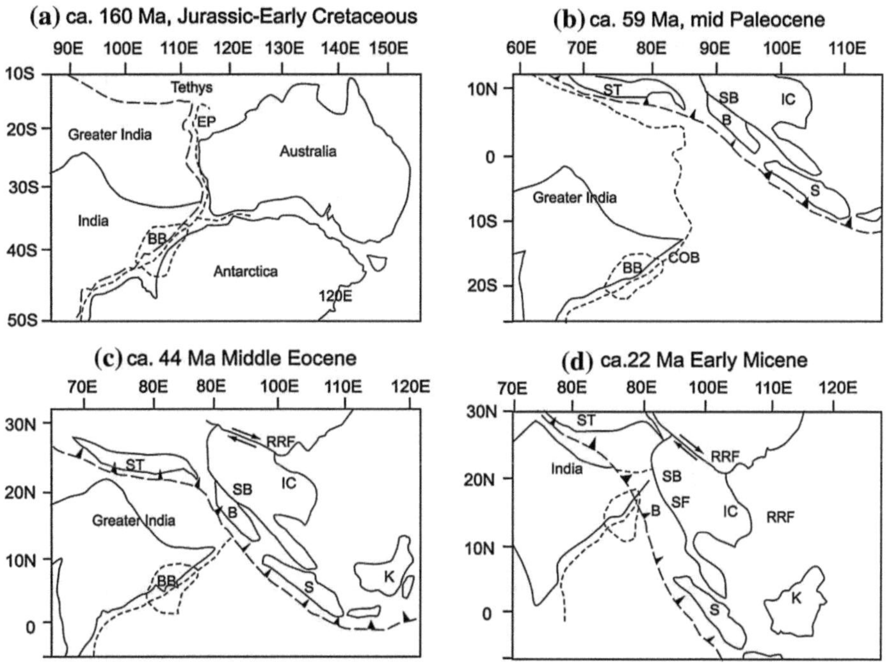

Fig. 4.2 Evolution of Bengal Basin since jurassic on the basis of plate tectonics (After Lee and Lawver, 1995)

In the mean-time subduction of the Indian plate beneath the Asian plate created a sutured zone of plate zone in the northern Himalayas while in the east the Indo-Burman range marks the zone of plate and the Asian plate continues and this is evidenced from the release of stress in the form of earthquakes in the vicinity of the plate boundary from time to time (Fig. 4.2).

Hence, the geological evolution of Bangladesh is basically related to the uplift of the Himalayan Mountains and outbuilding of deltaic landmass by major river systems originating in the uplifted Himalayan. Only the eastern part of Bangladesh has been uplifted into hilly landform incorporating itself into the frontal belt of the Iindo-Barma range lying to the east.

4.2 Delta Formation and Landform Development in Response to Sea Level Changes During Quaternary Period

4.2.1 Early Development: 18000–7000 cal year BP

The Bengal Basin comprised incised alluvial valleys and exposed lateritic uplands during the last sea-level low stand; at least 100 m or much lower than the present

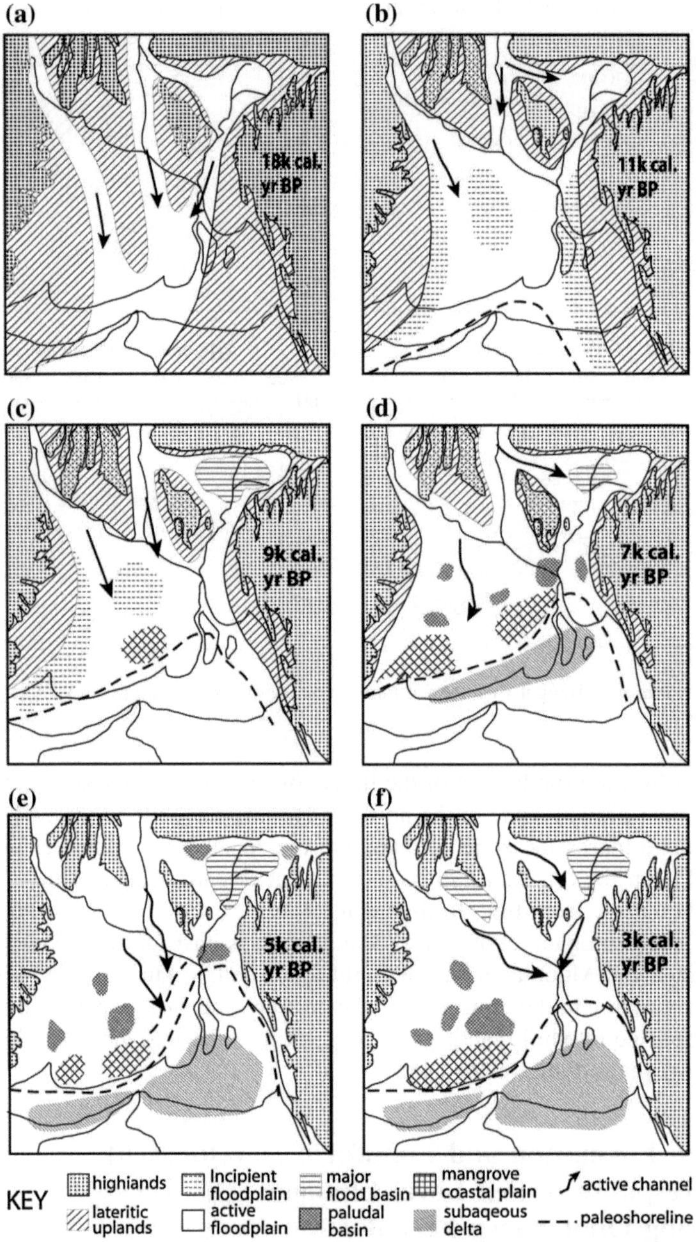

Fig. 4.3 Delta formation during quaternary period. The evolution of Bangladesh since 18000 years BP A-F (after Goodbred and Kuehl 2000)

level (Umitsu 1993) (Fig. 4.3), when river discharge was very low and perhaps negligible compared with modern output (Cullen 1981; Milliman 1983). During that time, the altitude of the Quaternary sedimentary contact in the northern part of the present Ganges Delta was found lower than 70 m (Umitsu 1993). Therefore, during the maximum epoch of the last glacial age, the rivers flowing in the Ganges Delta and its surrounding region dissected former surfaces, and deposited sandy gravels in the valley (Umitsu 1993). It was not until, 15000 years BP that significant sediment input is recorded on the upper Bengal Fan (Weber et al. 1997), apparently indicating climatic warming and increased precipitation in the Himalayas.

During 12000 years BP, the sedimentary condition and landforms near Khulna suggests a floodplain condition. Sedimentary facies showed that Khulna was similar to those of the present Bhahmaputra–Jamuna river bed during that time. During a period between 12000 and 10000 years BP, the surface of the delta and floodplain was slightly dissected, and shallow valley plains seem to have been formed along the rivers in Ganges Delta. This slight dissection may be caused by a temporal regression of sea-level (Umitsu 1993).

By, 11500 years BP, intensity of the humid southwest monsoon had grown to stronger-than-present levels (Gasse et al. 1991) and forced tremendous river sediment discharge of, 2:5 £ 109 t = year (Goodbred and Kuehl 2000). In the Bengal Basin at this time, accommodation and stream gradient adjustment favored infilling of the alluvial valleys with the Sand facies, but peak sedimentation rates on the fan (Weber et al. 1997) indicate limited capacity of the incipient delta to trap much of this immense discharge. By 10000–11000 years BP a major transition had occurred in the southern Bengal Basin when the finegrained Lower Delta Mud was widely deposited over the lowstand Oxidized and alluvial Sand units.

Eustatic sea level had risen to 245 m by 11500 years BP (Fairbanks 1989; Blanchon and Shaw 1995) and intersected a large portion of the low stand basin surface, causing back-flooding, extensive sediment trapping, and initial establishment of the G–B delta system (Fig. 4.3). Timing of this shift also corresponds with a very sharp decrease in fan sedimentation after 11000 years BP (Weber et al. 1997) and indicates rapid onshore migration of the major depocenter. The Lower Delta Mud facies, representing intertidal, mangrove, and nearshore environments, dominated strata deposition in this newly formed coastal plain. This setting persisted with little apparent change for the next 2000–4000 years and indicates relative shoreline stability during this period in which sea-level rose 30–35 m at a mean rate of 1 cm/year (Fig. 4.3). In addition to sediments trapped at the shoreline during the early Holocene, major fluvial and flood basin fill sequences were accreting throughout the upper Bengal Basin. Sand sequences continued to be deposited in the central alluvial valleys, as well as sequestered to the backflooded Sylhet Basin from 11000 to 9000 years BP. Fine-grained Sylhet Muds blanketed the Sylhet after 9000 years BP and suggest that the Brahmaputra flowed along its western course from this time until, 7500 years BP, discharging the majority of its load to the coast.

4.2.2 *Late Development: 7000 cal year BP to Present*

When the rate of sea-level rise slowed, 7000 years BP, maximum transgression was reached in the western Bengal Basin (Banerjee and Sen 1988; Umitsu 1993) and the delta changed from an aggradational to progradational system (Fig. 4.3).

The main fluvial depocenter migrated sea ward, more of the river's suspended load was forced onto the shelf, leading to growth of a muddy submarine delta by 7500 years BP (Kuehl et al. 1997). This marks initial development of the modern compound delta system with prograding subaerial and subaqueous clinoforms. Concurrent with delta growth on the shelf, shoreline progradation continued in the western Bengal Basin and approached its present extent by 5000 years BP (Fig. 4.3). Coastal peats formed in the Thin Mud facies after this time and indicate abandonment and eastward migration of the active Ganges distributaries (cf. MPO 1985; Banerjee and Sen 1988). In the Sylhet Basin, very rapid infilling (0.2 cm/year) ensued from 7500 to 6000 years BP when the Brahmaputra river switched to its eastern course and drained into this large, deepened basin. Broad Brahmaputra–Jamuna floodplains formed and at Kishorganj in the Sylhet basin shows marshy areas during this time (Umitsu 1993). On the eastern coastal plain south of the Sylhet, local peats indicate limited clastic input to this area from 7400 to 6000 years BP (Brammer 1996). Extensive shallow (7 m) estuarine mud deposits in this region also suggest limited delta growth at this time (Brammer and Brinkman 1977).

Between 5000 and 6000 years BP, the Brahmaputra appears to have switched back to its western course (Fig. 4.3), indicated by renewed sand deposition along this former channel and deposition of a silt layer in the western Sylhet. Also, peat horizons in the eastern and western Sylhet correspond to this time (4719 and 4207 years BP, respectively) and support a lack of significant sediment input (Umitsu 1993). With this Brahmaputra course change and eastward migration of the Ganges, the two rivers may have first joined in the Holocene near their present courses (Fig. 4.3). Delta growth after 5000 years BP largely occurred in east Bengal near the modern river mouth estuary, prograding seaward through the present. At least one other Brahmaputra avulsion cycle occurred, depositing the upper sand sequence in the Sylhet before changing to its western course 150 years ago (Fergusson 1863). With limited accommodation in the Sylhet during the latest Holocene, most of the load is presumed to have bypassed to the coast. Offshore, the subaqueous delta prograded continuously since 7500 years BP, with base of the foresets reaching the 80 m isobath at present (Kuehl et al. 1997; Michels et al. 1998).

The lack of major sediment input in the east contrasts with progradation of the western Bengal coastline, and we presume that much of the Brahmaputra load was sequestered to the Sylhet Basin causing maximum transgression to occur 1000–2000 years later in the east. At that time Khulna was close to the coastline where mangrove trees grew widely. And Khulna was under a relatively strong marine influence, and the coastline of the Ganges Delta retreated slightly to the north of the present Khulna city. The coastline advanced slightly towards the Bay of Bengal during that time, when broad marshy peat lands spread out in the central Ganges Delta (Umitsu 1993).

During 3000 years BP, the transgression rate became slower in the late Holocene, and peaty and marshy lands spread out widely in the Ganges Delta and its surrounding region, and that drainage conditions in the lowland became worse. Along the main channels such as the Ganges and the Brahmaputra rivers, however, distinctive sedimentary and landform changes have continued.

Presently, with about half of its surface below the 10 m contour line, Bangladesh is located at the lowermost reaches of three mighty river systems—the Ganges–Padma river system, Brahmaputra–Jamuna river system and Surma–Meghna river system.

4.2.3 Physiographic Regions in Bangladesh

Bangladesh has four distinct physiographic regions each having distinguishing characteristics of its own.

1. The north-eastern hilly region
2. The great Table land
3. Flood plains
4. Delta

The Eastern and Northern Frontier Hilly Regions comprising the extensive eastern hilly regions, hills of Lalmai and north-eastern Sylhet district and a narrow strip of a series of low hill ranges and isolated circular and elongate hillocks separated by recent alluvium along the northern frontier of districts of Sylhet and Mymensingh.

The great Table Land extends from the Lalmai Hills of Comilla district and the marginal low hills of the districts of Chittagong and Sylhet in the east through Dhaka and Rajshahi Divisions to West Bengal of India. The river systems of the Meghna and the Jamuna trisect the Table Land giving rise to three large blocks of high lands that exhibit smooth rolling topography. Therefore, at the end of the last glaciation, amplified monsoon water plus deglaciated melt water from Himalayas enormously flowed through the narrow Paleo river systems which were overloaded and overflowed. Due to the strong hydrodynamic condition, the initial Barind and Madhupur surfaces were highly dissected, created some pools and depressions, left a number of north–south elongated reddish-brown islands or terraces (Monsur 1995).

The Table Land covers an area of about 10 % of Bangladesh. The Madhupur Tract measures about 4105 sq. km. The elevation of the Table, which occupy higher ground than the flood plain, varies from place to place. The Table Land attains heights of 100 feet (about 33 m) at Lalmai, 50 feet (about 16 m) in Madhupur tract. The average elevation of this land is most likely to be more than 50 feet above msl.

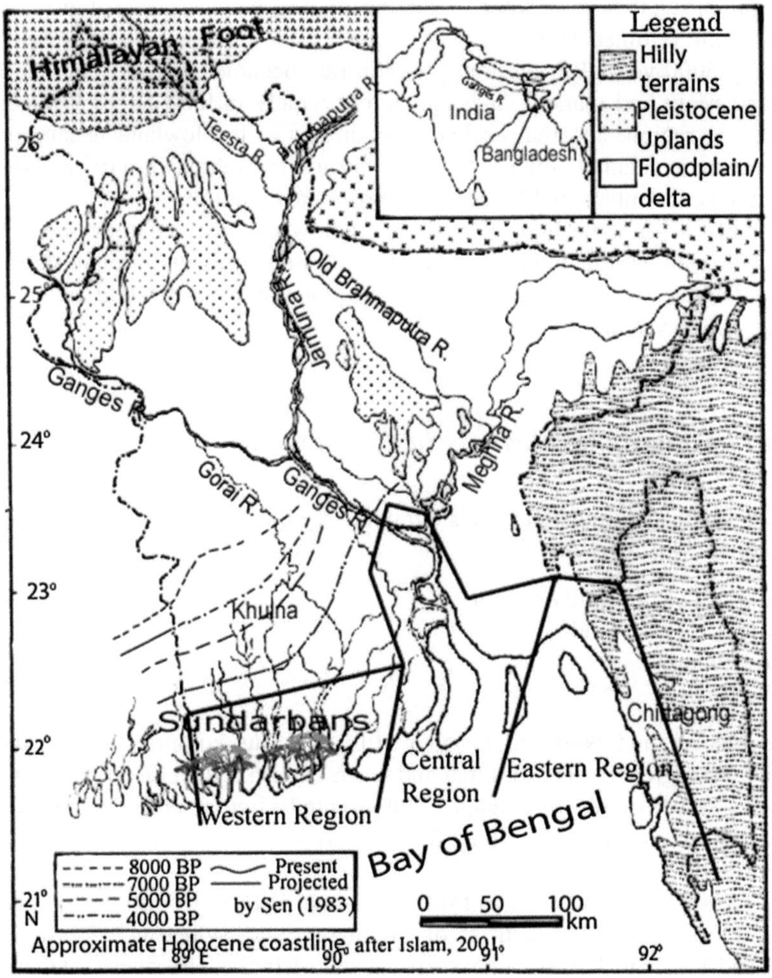

Fig. 4.4 Physiography of Bangladesh

The flood plains of the Ganges, the Brahmaputra, and the Meghna cover approximately 40 % of Bangladesh. The Jamuna and the old Brahmaputra with their main tributary the Tista and a good number of small tributaries constitute the largest floodplain of Bangladesh. Excepting for a small portion in the north, the entire flood plain lies within the Table Land region. The flood plain and the Table Land almost completely cover two Divisions, namely, Rajshahi and Dhaka. The elevation of the major part of the floodplains of these three great rivers ranges from 10 to 45 feet (3–15 m) above msl.

The Ganges builds up the largest area of the delta. The Ganges delta includes several districts of West Bengal and the districts of Kushtia, Khulna, Jessore and western part of the districts of Faridpur and Barisal. The elevation of the deltaic region varies from 1 to 8 m above msl (Fig. 4.4).

Bangladesh, a flood plain delta, is a land of rivers and canals. The country is sloping gently from the north to the south, meeting the Bay of Bengal at the southern end. The whole coast runs parallel to the Bay of Bengal, forming 710 km long coastline (CZPo 2005). According to the coastal zone policy (CZPo) of the Government of Bangladesh, 19 districts out of 64 are in the coastal zone covering a total of 147 upazillas of the country. Out of these 19 districts, only 12 districts meet the sea or lower estuary directly. The country has three distinct coastal regions, namely the western, central and eastern regions. The east and west parts of the coast are of very different characters. The east coast is very much unbroken and straight and is only intercepted by smaller rivers like Karnafuli, Sangu, Matamuhuri Bogkhali etc. whereas remaining coast of Gangetic delta is very much broken and intercepted by large estuaries like arms of sea carrying sediment laden water. The east coast has a narrow coastal plain hardly a few km wide that quickly goes up into the hills while Gangetic delta in Bangladesh is very flat and tide travels far inland over 150 km into the country (Matin 2009).

The zone is divided into exposed and interior coast according to the position of land. The upazillas that face the coast or river estuary are treated as exposed coastal zone. The exposed coast embraces the sea directly and is subject to be affected highly by the anticipated sea level rise. The coastal zone covers 47201 sq. km land area, which is 32 % of total landmass of the country (Islam 2004).

The southern part of Bangladesh falls under coastal zone that receives discharge of numerous rivers, including Ganges-Brahmputra-Meghna (GBM) river system, creating one of the most productive ecosystems of the world. Except Chittagong-Cox's Bazar, all parts of the coastal zone are plain land with extensive river networks and accreted land, which is known in Bangladesh as char land.

References

Banerjee M, Sen PK (1988) Palaeobiology and environment of deposition of Holocene sediments of the Bengal Basin, India. In: Proceedings of the second conference on the Palaeoenvironment of East Asia from the mid-tertiary, vol 1. Geology, Sea-level Changes, Palaeoclimatology and Palaeobotany, Centre of Asian Studies, Hongkong, pp 703–730

Banglapedia (2003) National Encyclopedia of Bangladesh. Asiatic Society of Bangladesh, Dhaka

Blanchon P, Shaw J (1995) Reef drowning during the last glaciation: evidence for catastrophic Sea-level rise and ice sheet collapse. Geology 23:4–8

Brammer HB (1996) The geography of the soils of Bangladesh. University Press Ltd., Dhaka

Brammer H, Brinkman R (1977) Surface-water gley soils in Bangladesh: environment, landforms and soil morphology. Geoderma 17:19–109

Cullen JL (1981) Microfossil evidence for changing salinity patterns in the Bay of Bengal over the last 20,000 years. Palaeogeogr Palaeoclimatol Palaeoecol 35:315–356

CZPo (2005) Coastal Zone Policy, Ministry of Water Resources, Government of the People's Republic of Bangladesh, Dhaka

Fairbanks RG (1989) A 17000 year glacio-eustatic sea level record: influence of glacial melting rates on the Younger Dryas event and deep ocean circulation. Nature 342:637–642

Fergusson J (1863) Delta of the Ganges. Quat J Geol Soc Lond XIX:321–354

Gasse F, Arnold M, Fontes JC, Fort M, Gibert E, Huc A, Li BY, Lju Q, Melieres F, Vancampo E, Wang FB, Zhang QS (1991) A 13,000-year climate record from western Tibat. Nature 353:742–745

Goodbred SL, Kuehl SA (2000) The significance of large sediment supply, active tectonism, and eustasy on margin sequence development: late Quaternary stratigraphy and evolution of the Ganges-Brahmaputra Delta. Sed Geol 133:227–248

Islam MR (ed) (2004) Where Land Meets the Sea: A Profile of the Coastal Zone of Bangladesh, The University Press Limited, Dhaka

Kuehl SA, Levy BM, Moore WS, Allison MA (1997) Subaqueous delta of the Ganges-Brahmaputra river system. Marine Geol 144:81–96

Matin MA (2009) Mean Sea level variation in the Bangladesh Coast. Unpublished article

Michels KH, Kudrass HR, Hübscher C, Suckow A, Wiedicke M (1998) The submarine delta of the Ganges-Brahmaputra: cyclone-dominated sedimentation patterns. Mar Geol 149:133–154

Milliman JD, Mead RH (1983) World-wide delivery of river sediment to the oceans. J Geol 91:1–22

Monsur MH (1995) An introduction to the Quaternary geology of Bangladesh. City Press and Publications, Dhaka

MPO (1985) Geology of Bangladesh, MPO Technical Report no 4. Master Plan Organization. Ministry of Irrigation, Water Development and Flood Control, Government of Bangladesh, Dhaka

Umitsu M (1993) Late quaternary sedimentary environments and landforms in the Ganges Delta. Sed Geol 83:177–186

Weber ME, Niessen F, Kuhn G, Wiedicke-Hombach M (1997) Calibration and application of marine sedimentary physical properties using a multi-sensor core logger. Mar Geol 136(3–4):151–172

Chapter 5
Sea Level Changes Scenarios During Holocene in Bangladesh

Paleo mean sea levels (MSL) are reconstructed from a wide variety of environmental indicators, including morphological and archaeological data, palaeosols and other lithostratigraphic changes. Litho-bio-chrono-stratigraphic techniques to reconstruct the Holocene sea level changes are well established and have been applied successfully in many coastal areas of the world (Tooley 1978; Shennan 1982; Ireland 1987; Islam 2001a, b). In this study, lithofacies characteristics, *mangal* peat, and paleophysiographic evidence have been considered to indicate the past position (space) of the relative sea level.

In coastal areas, mangrove peat formation is favoured by relative sea-level rise, but this rise must not be too slow or too rapid (Einsele 2000), which implies that peat beds are formed under stable sea-level conditions (Umitsu 1993). To construct the Holocene RSL curve in Bangladesh, a total of eleven index samples were used. Their radiocarbon ages, samples types and altitudes relative to the MSL are summarised in Table 5.1. The paleo-MSL altitudes have been determined based on indicative meaning for elevation (occurrence range) of the *mangal* peat of the sample areas. Considering the paleo-environmental records of each site, the relative sea-level changes during the Holocene can be divided into 5 phases (Figs. 5.1 and 5.2), which are described below:

5.1 Phase 1 (Approximately 7500 years BP)

The sand flat facies reveals a marine transgression of the bay at approximately 7500 years BP, which submerged the Sony site with seawater (Rashid et al. 2009). Due to the rising of the sea level, the mangrove forests did not maintain their habitat. The sand flat sequence contains abundant mangrove pollens at 3.75 m below the present MSL, which reveals that mangrove vegetation was developed very close to the succession site (Rashid et al 2013). At this stage, the relative sea level appears to be considerably higher, which suggests that it had reached the nearby elevated land, where mangroves forest might have been growing (Fig. 5.1). During this mid-Holocene highstand, the shoreline moved inland, and the area was

T. Rashid, *Holocene Sea-level Scenarios in Bangladesh*,
SpringerBriefs in Oceanography, DOI: 10.1007/978-981-4560-99-3_5,

Table 5.1 Summary of Sea-level index samples for the central part of Bengal Lowland

Location	Sample no	Radiocarbon age (cal BP)	Altitude from present MSL (m)	Sample identification	Sample type	Universal occurrence range (suggested [a]altitude of paleo-msl)
Sony	1	7570–7430	−3.75	Plant remains (wood)	Mangrove pollen in sand flat	MLWL-MWL (1−2 m higher than present msl)?
	2	6670–6410	−1.60	Peat	Mangrove peat in salt marsh	MWL ± 1 m (1−2 m lower than present msl)
	3	6670–6410	−1.6	Peat	Marine and brackish water diatom in salt marsh	Not known (1−2 m lower than present msl)
	4	Between 7500 and 6500	−2.8	Lithofacies	A sharp erosional contact	Not known
	5	4080–4030	+0.9	Peat	Freshwater peat	Above SL
	6	1820–1530	+1.8	Peat	Freshwater peat	Above SL
Dobadia	7	6280–5940	+4.5	Diatom, wood (not in situ)	Marine, brackish and freshwater diatom	MWL ± 1 m (4.5–5 m higher than present msl)
	8	5980–5700	+5.7	Peat	Freshwater peat	Above SL
Vatpara	9	5580–5300	+0.4	Peat	Mangrove peat	MWL ± 1 m (1−2 m higher than present msl)
Chatbari	10	5440–5410	+0.9	Diatom, Peat (not in situ)	Mangrove organic matter	Not known (1−2 m higher than present msl)
	11	4870–4780	+2.1	Peat	Freshwater peat	Above SL
	12	3080–2760	+3.1	Peat	Freshwater peat	Above SL
Nayanipara	13	3690–3440	−0.20	Peat	Freshwater peat	Above SL (0.3–0.5 m lower than present msl)

MLWL-mean low water level; MWL-mean water level; SL-sea level

[a] Altitude of paleo-msl could not be determined on the basis of present tidal levels as the study area is far away (200 km) from the present coastal region and as a result the area is free from present tidal range. The altitude of paleo-msl is suggested on the basis of indicative meaning of mangal peat and mangroves, diatoms as well as of paleo-physiographic evidences of the sample areas

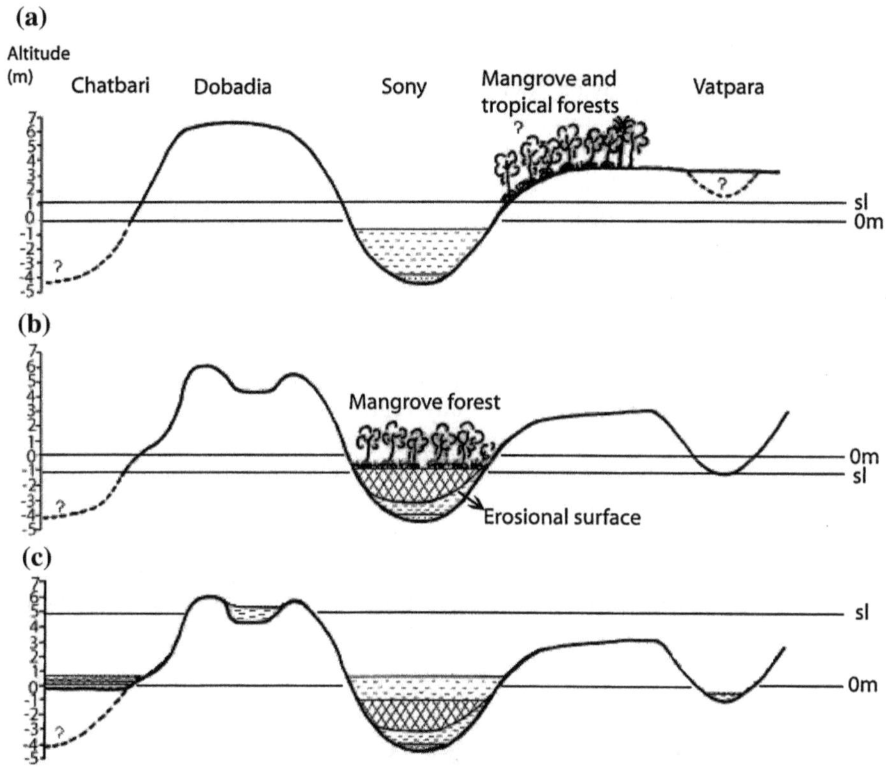

Fig. 5.1 Paleoenvironment of the study sites (phase 1 to phase 3). Schematic presentation of paleoenvironment in relation to sea-level changes in and around the sites (phase 1–3). **a** Transgression around 7500 cal BP. **b** Regression around 6500 cal BP **c** Transgression around 6000 cal BP

invaded with saltwater (Rashid et al 2013). The saltwater mixed with freshwater by tidal action and, thus, an estuarine (Pritchard 1967) environment was created in the central part of Bangladesh. The pollen results show that this saltwater intrusion affected the terrestrial freshwater plants, and the environment was conducive to the development of salt resistant mangrove forests in and around Dhaka City. The mud flat facies (SU2, VU1) around the area might have developed as a consequence of the transgression at Sony (Rashid et al. 2009, 2013).

In the first phase, a relative sea-level high stand occurred at approximately 7500 years BP (Fig. 5.1), although the height is yet to be determined (Sample no. 1 in Table 5.1). Because the transgression is observed at the low tide line intertidal coastal-plain at Sony and the location of mangrove ecosystem is unknown, the height of this high stand could not be determined.

5.2 Phase 2 (at Approximately 6500 years BP)

The salt marsh facies provides evidence of a regression of the bay at approximately 6500 years BP at Sony (Rashid et al 2013). Due to this regression, the mud flat deposit (SU2) was eroded and shaped by the upper bed cut, and the environment of Sony was gradually changed from a mud flat to a salt marsh (SU3). The salt marsh contains abundant mangrove pollens and marine and brackish water diatoms. The radiocarbon dating during 6670–6410 years BP is obtained from this unit (at −1.6 m) (Rashid et al 2013). Because the site had an inflow of shallow marine water in this regressive episode, the mangrove forests took their positions easily (Fig. 5.1), and, therefore, *mangal* peat formed between −2.8 and −1.1 m (Fig. 5.1) (Rashid et al 2013). With this regression, some erosive activities are thought to have occurred around the study region, such as in Dobadia, Sony and Vatpara. Besides palynology, lithofacies analysis indicates that the sites were affected by erosion activities, which strongly support this regression (Fig. 5.1). The sharp contact at −2.8 m MSL in a salt marsh and the *mangal* peat between −1.1 and −2.8 m MSL reveal that the relative sea level was considerably lower, perhaps by 1–2 m, than the present MSL (Sample nos. 2–4 in Table 5.1) (Fig. 5.1) (Rashid et al 2013). The paleo-coastline, then, may have moved slightly to the south.

5.3 Phase 3 (Approximately 6000 years BP)

The mud flat facies and radiocarbon dating results of the Dobadia site (DU1) show a transgression of the bay at approximately 6280–5940 years BP (Rashid et al 2013). After the regression, the relative sea level started to rise and sediment accumulation occurred. Subsequently, the valley was filled with 1.5 m thick marine mud and the secondary succession of Madhupur red clay in response to the transgression. During this sea-level rise, the Sony site was drowned by seawater (Fig. 5.1).

The burrows with marine mud indicate a rapid transgression around the site, and the diatom results determine the height of this high stand, which was at least 4.5 m (Fig. 5.1) above the modern sea level (sample nos. 7 and 8 in Table 5.1) (Fig. 5.1). The curve shows a sharp transgression at approximately 6000 years BP after the regressive phase (Rashid et al 2013). After this short phase of transgression, the relative sea level started to fall quickly, and consequently, freshwater peat was developed at the site at approximately 5980–5700 years BP (Rashid et al 2013).

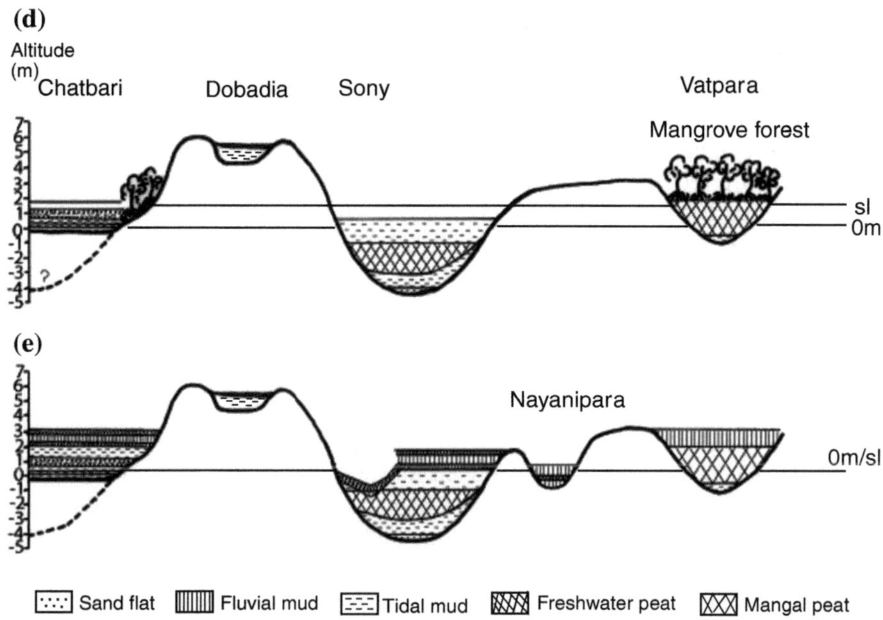

Fig. 5.2 Paleoenvironment of the study sites (phase 4 to phase 6). Schematic presentation of paleoenvironment in relation to sea-level changes in and around the sites (phase 4–5). **d** Regression around 5500 cal BP **e** Regression between around 5000 and 1500 cal BP

5.4 Phase 4 (Between Approximately 5500 and 5000 years BP)

The salt marsh facies provides evidence of a regressive episode at approximately 5580–5300 years BP at the Vatpara site after the transgression observed at Dobadia (Rashid et al 2013). The fact that there are a significant number of mangrove pollens at Vatpara reveals that the site had an inflow of shallow marine water during this regressive episode. During this regression the Sony site was submerged by seawater, and mangroves disappeared; consequently, a mud flat succession (SU4) was developed in an intertidal coastal plain. With this phase, the salt marsh observed at Sony regressed and the mangrove forests kept their positions at Vatpara, and hence, the *mangal* peat developed between 0 and 2 m above MSL (Fig. 5.2) (Rashid et al 2013). The Vatpara site is approximately 18 km north–east of Sony. Because the Vatpara site is slightly elevated, it can be inferred that when Sony was submerged by the sea (consequently, developing a mud flat during this episode), Vatpara could not have been drowned and received saltwater thus creating a mangal ecosystem (Fig. 5.2) in intertidal coastal plain. The abundant mangrove pollen at 0.4 m above MSL at Vatpara indicates the marine environment around the site (Rashid et al 2013).

However, the lagoonal estuarine mud facies shows that the Chatbari site also experienced a marine environment at approximately 5500 years BP (Rashid et al 2013). The few mangrove palm pollens and a significant number of marine diatoms at 0.9 m above MSL imply that the sea level was slightly higher than its depositional height: at least 1 m higher than it is today (Rashid et al 2013).

It can be inferred that the relative sea level higher between approximately 5500 and 5000 years BP and that the height for this phase seems to be 1–2 m greater than the modern sea level (Sample nos. 9 and 10 in Table 5.1).

5.5 Phase 5 (Between Approximately 5000 and 1500 years BP)

The freshwater peat sequences developed during 4870–4780 (+2.1 m MSL) and 3080–2760 (+3.1 m MSL) years BP at Chatbari, 4080–4040 (+0.9 m MSL) and 1820–1530 (+1.8 m MSL) years BP at Sony and 3690–3440 (−0.2 m MSL) years BP at Nayanipara strongly support that the facies were developed under the regressive condition of the bay and that the area was surrounded by a freshwater terrestrial environment due to the relative sea level starting to fall after the transgression of the bay (Rashid et al 2013). The fluvial facies indicates that the relative sea level started to fall after the transgression between approximately 5500 and 5000 years BP. These results reveal that at approximately 4870–4780 years BP and onwards, the sites were dominated by a freshwater environment and that the *mangal* ecosystem had been replaced by freshwater vegetation due to the relative sea-level fall. Considering the litho-biofacies, the results also indicate that the Nayanipara site, located in the Madhupur tract, was eroded away with a regression, the relative sea level dropped below the m.s.l. during 3690–3440 years BP, and the freshwater peat accumulated (from −0.3 to 0 m) just below the present sea level (Figs. 5.2 and 5.3) (Rashid et al 2013). With the sea-level fall, the osional activities ended and the erosional surface was filled by the Holocene freshwater deposits. The sharp boundary with the fresh water marshy land facies at Sony also supports an erosional phase before 4080–4040 years BP (Rashid et al 2013). It is, however, inferred that the relative sea-level started to fall at approximately 5000 years BP, which coincides with the weaker ISM from 5000–1200 years BP, particularly with a dry event at ca. 3600 years BP (Fleitmann et al. 2003; Van Campo et al. 1996; Wang et al. 2005; Bookhagen et al. 2006).

From the RSL curve (Fig. 5.3), it is unequivocally stated that the relative sea level started to fall at approximately 5000 years BP (Sample nos. 5, 6, 11, 12, and 13 in Table 5.1). The curve shows a lower relative sea level at approximately 3690–3440 years BP, and the relative sea level may have dropped between −0.3 and −0.6 m (Fig. 5.3) (Rashid et al 2013). Thereafter, the relative sea level started to move towards the present position and has not noticeably migrated towards land.

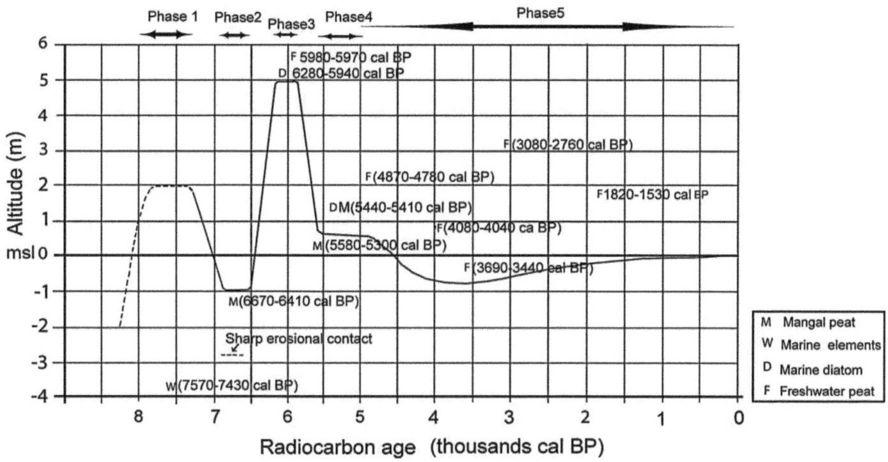

Fig. 5.3 Relative sea level curve of Bangladesh during Holocene

5.6 Paleo-Coastline of Bangladesh During Early to Middle Holocene

On the basis of marine transgression and regression phases an approximate paleo-coastline has been drawn during the Holocene. The study shows that the sand flat facies developed in the low tide line of the intertidal coastal plain which received mangrove pollen from nearby mangrove forest which was elevated than Sony site at approximately 7500 years BP. The Vatpara was 18 km (behind) north–east and it appears that the mangrove forest was available in between Sony and Vatpara. The mangrove forest was mixed with tropical forest elements imply that the vegetation coverage might have been developed in a supratidal coastal plain- at marine limit (break in terrestrial vegetation). Therefore, it may be speculated that the paleo-coastline was very close to the site Sony during 7500 years BP. During the regressive phase, both the site Vatpara and Dobadia might have eroded which were much higher than the surrounding area but the site Sony still received marine water and the paleo-coastline might retreated 1–2 m south from the previous position. Thereafter, due to the transgression at approximately 6000 years BP at Dobadia site the coastline moved again towards north and was very close to Dobadia site.

During 5500 years BP, Vatpara, Sony and Chatbari sites were also developed in intertidal coastal plain. It implies that the coastline at approximately 5500 years BP was very close to the coastline during 7500 years BP (Fig. 5.4). It indicates that both peaks were attained at same hight: 1–2 m higher than the present MSL. From Islam (2001a, b) and Umitsu (1987, 1993) it is clear that during mid Holocene period the coastline was slightly north of Khulna city and was possibly very close to the sites of Jessore and Narail. Islam (2001a, b), however, pointed

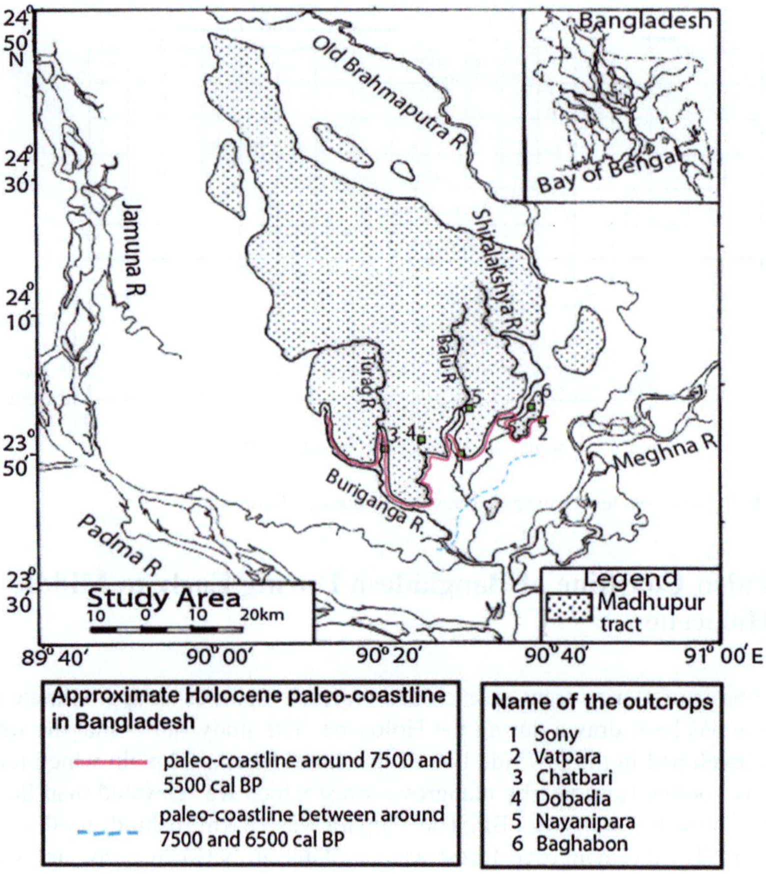

Fig. 5.4 Paleo coastline of Bangladesh during Holocene

that due to lack of evidence it is difficult to extend the lines beyond the Padma-Meghna course although records from Matuail suggest that the site was under a marine influence until about 2400 cal. years BP. The paleo-coastlines (Fig. 5.4) reveal that during the last 7500 years, the central part of Bangladesh, beyond the present day Padma River course, has been subjected to a submergence phase which lasted until 5500 cal BP, followed by a drop in sea level from −1 to −2 m between around 7500 and 6500 cal BP.

Last but not the least, Towhida and Zahid (2011) found that the Shutrapur site, located in the western fringe of Madhupur tract which is 23 km south from the Chatbari site, could not receive marine water while the Chatbari site experienced with marine environment at approximately 5500 years BP. It infers that the marine limit persisted at or around the Chatbari site and marine influence could not reach beyond that.

References

Bookhagen B, Fleitmann D, Nishiizumi K, Strecker MR, Thiede RC (2006) Holocene monsoonal dynamics and fluvial terrace formation in the northwest Himalaya, India. Geology 34:601–604

Einsele Gerhard (2000) Sedimentary basins: evolution, facies, and sediment budget. Springer, Germany

Fleitmann D, Burns SJ, Mudelsee M, Neff U, Kramers J, Mangini A, Matter A (2003) Holocene forcing of the Indian monsoon recorded in a stalagmite from southern Oman. Science 300:1737–1739

Ireland S (1987) The Holocene sedimentary history of the coastal lagoon of Rio de Janiero state, Brazil. In: Tooley MJ, Shennan I (eds) Sea-level changes. Blackwell, Oxford

Islam MK (2001a) Evidences of neotectonics in and around Lalmai Hills, Bangladesh. Bangladesh J Geol 20:55–66

Islam MS (2001b) Sea-level changes in Bangladesh: the last ten thousand years. Asiatic Soc Bangladesh, Dhaka

Pritchard DW (1967) What is an Estuary? Physical viewpoint. In: Lauff GD (ed) Estuaries. American Association for the Advancement of Science, Washington DC, pp 3–5

Rashid T, Hassan Z (2011) Paleoenvironment reconstruction of Madhur tract. Unpublished M.Sc thesis, University of Dhaka, Bangladesh

Rashid T, Monsur MH, Suzuki S, OOI N (2009) Reconstruction of Holocene Paleoenvironment and evidence of sea-level changes in the Bengal Lowland. J Geogr 118(4):631–645

Rashid T, Suzuki S, Sato H, Monsur MH, Saha SK (2013) Relative sea-level changes during the Holocene in Bangladesh. J Asian Earth Sci 64:136–150

Shennan I (1982) Interpretation of Flandrian sea-level data from the Fenland. England Proc Geol Asso 83(1):53–63

Tooley MJ (1978) Sea-level changes during the last 9000 years in North-west England. Geogr J 140:18–42

Umitsu M (1987) Late Quaternary sedimentary environment and landform evolution in the Bengal Lowland. Geogr Rev Jpn 60:164–178

Umitsu M (1993) Late Quaternary sedimentary environments and landforms in the Ganges Delta. Sed Geol 83:177–186

Van Campo E, Cour P, Sixuan H (1996) Holocene environmental changes in Bangong Cobasin (western Tibet), part 2: the pollen record. Palaeogeogr Palaeoclimatol Palaeoecol 120:49–63

Wang YJ, Cheng H, Edwards RL, He YQ, Kong XG, An ZS, Wu JY, Kelly MJ, Dykoski CA, Li XD (2005) The Holocene Asian Monsoon: links to solar changes and North Altantic climate. Science 308:854–857

Chapter 6
Discussions and Conclusion

The eustatic sea level was higher during the mid-Holocene in a considerable part of the world (Ingolfsson and Hjort 1999), including South America, Africa, Antarctica, Australia, and the Pacific and Indian Oceans (Isla 1989). Although the nature of the Holocene relative sea-level change is broadly similar in all locations of the Indo-Pacific region, differences do exist in the timing and magnitude of the mid-Holocene high stand across the region (Woodroffe and Horton 2005). Fixed biological indicators (oyster beds) have been used at one site on the coast of the South China Sea to infer that the amplitude and timing of the mid-Holocene high stand was no greater than 2 m above the modern sea level at 5140 years BP (Yim and Huang 2002). The Changjiang delta in China experienced the highest sea level, approximately 2–3 m higher than that at present, during approximately 6000 years BP (Hori and Saito 2000). Banerjee (2000) studied the regional RSL change on the East Indian coastline, utilising beach ridges and exposed *Porites* colonies to indicate two mid-late Holocene highstands, the first peaking at 7300 years BP, followed by an RSL fall, and a second pulse of minor RSL rise culminating between 4300 and 2500 years BP. From the sediment structure, mineralogy, and organic matter content, supplemented by radiocarbon dates, Hameed et al. (2006) contributed new evidence regarding sea-level change along the east coast of India indicating that the sea level rose slowly from 9000 to 1115 years BP. However, Long (2001) concluded that the RSL changed rapidly between 7800 and 4400 years BP because of a fall in meltwater discharge from the dwindling ice sheets of the northern hemisphere.

In this study, mangrove peat was used as a sea-level indicator to determine the magnitude of the paleo-MSL, as it is directly related to the position of the sea at the time of it deposition. Facies analysis was conducted to understand the sedimentary environment, and the altitude was measured to determine the paleo-MSL in Bangladesh during the Holocene. The study provides evidence of a mid-Holocene relative sea-level high stand in Bangladesh at approximately 7500 years BP. This high stand invaded the central part of Bangladesh, and paleo-coastline migrated north up to the Dhaka region, beyond the present day Padma River course. After the transgression, a regression occurred between approximately 6500 and 7000 years BP, which coincides with a prominent weak ISM and a dry phase

T. Rashid, *Holocene Sea-level Scenarios in Bangladesh*,
SpringerBriefs in Oceanography, DOI: 10.1007/978-981-4560-99-3_6,

between ca. 6700 and 6100 years BP (Fleitmann et al. 2003; Van Campo et al. 1996; Wang et al. 2005; Bookhagen et al. 2006). The sharp contact at −2.8 m MSL in a salt marsh and the *mangal* peat showed a considerably lower MSL, perhaps 1–2 m, than the present MSL, which is a truly unique finding of this study.

The present study showed a relative sea-level rise during 6280–5940 years BP cal BP at Dobadia which is located at the southern part of Madhupur tract in the central part of Bangladesh. The marine and bay diatoms found at 4.7 m at Dobadia is strongly supporting a marine environment which reveals that the site was submerged by sea water and the relative sea-level was about 4.7 m higher than the present mean sea-level in response to the transgression around 6280–5940 years BP which is thus far the highest level identified in Bangladesh. The Changjiang delta in China experienced the highest sea-level, approximately 2–3 m higher than that at present, during about 6000 years BP (Hori and Saito 2000).

The transgression event found at Dobadia around 6000 years BP seems that after the first peaking on the east Indian coastline at 7300 years BP (Banerjee and Sen 2000) and Bangladesh during 7500 years BP (Rashid et al. 2009), followed by a regression between around 7000 and 6500 years BP observed at Sony (Rashid et al. 2009), the sea-level rose again with 4.5 m peaking. After a short span, it can be inferred from the fresh water pollen **results** that the relative sea-level started to fall rapidly and the site was surrounded with fluvial environment around 5980–5970 years BP. From sediment structure, mineralogy, organic matter content, supplemented by radiocarbon dates, Hameed et al. (2006) added new evidence, regarding the sea-level change along the east-coast of India, that the sea-level rose slowly since 9000–1115 years BP. On the other hand, Long (2001) concluded that RSL changed rapidly between 7800 and 4400 years BP because of a fall in meltwater discharge from the dwindling ice sheets of the northern hemisphere. Islam (2001) showed a transgression event during 5850–4580 years BP at Matuail which is about 20 km south from Dobadia. As Dobadia is much higher compared to Sony and Matuail the environment around Dobadia had no marine influence with the falling stage of sea-level whereas Matuail was still under marine influence.

During the regressive phase between around 7000 and 6500 years BP observed at Sony, the paleo-MSL was lower than the present MSL (Rashid et al. 2009). The transgression at Dobadia seems to be very rapid and abrupt as it rose about 4.5 m just after the regression around 7000 and 6500 years BP. In response to this abrupt transgression, the valley filled with 1.5 m thick marine mud. The depositional environment, thereby, is indicating that after the regression found at Sony (Rashid et al. 2009), relative sea-level started to rise at about 6000 years BP very sharply and abruptly which was about 4.5 m higher than modern MSL. And after a short span the relative sea-level started to fall rapidly and the site was surrounded with fluvial environment around 5980–5970 years BP.

Another regression was found after 6000 years BP, at approximately 5500 years BP, although the sea level was 1–2 m higher than the present sea level. Our study, for the first time in Bangladesh, has determined the height of the RSL high stand.

The results show that after 5000 years BP, the sea level dropped and brackish water mangrove species were replaced by fresh water vegetation at our study sites. The results also suggest freshwater marshy land before 1820–1530 years BP, as there was no marine influence in this region. This study also identified another RSL drop event at approximately 3690–3440 years BP, when the RSL fell more than 0.3 m below the present MSL, which coincides with a dry event at 3600 years BP (Fleitmann et al. 2003; Van Campo et al. 1996; Wang et al. 2005; Bookhagen et al. 2006). The RSL curve is also strongly supported by paleo-climatic evidence during the Holocene.

The present study also provides strong evidence of brackish water intrusion in the central part of Bangladesh between approximately 7500 and 5500 years BP by the rising level of the bay. As mangrove swamps are easily recognized habitats along brackish estuaries, the study area experienced brackish water and mangrove vegetations at approximately 7570–7400, 6670–6500, 6280–5940, 5580–5300, and 5440–5410 years BP. The study also showed that with the fall of the sea level, from 4870–4780 years BP and onwards, the brackish water did not further invade towards the central part of Bangladesh. The mangrove forests disappeared, and therefore, freshwater peat layers were accumulated at Chatbari during 4870–4780, 3080–2760 years BP, at Sony during 4080–4030 and 1820–1530 years BP and at Nayanipara during 3690–3440 years BP. Hence, it can be unequivocally stated that the sea level started to retreat towards the south before 4870–4780 years BP. After this period, the brackish water environment could not have prevailed around the study region and was replaced by a freshwater environment.

All paleoenvironmental records including litho-biofacies and paleosols of the section indicate that the Pleistocene-Holocene erosional valley of Madhupur tract (Monsur 1995; Rashid et al. 2009) had eroded and inundated with brackish water in response to a mid-Holocene high stand which was about 4.5 m higher than the present MSL. Tree trunks, woods, and roots of mangroves are available in the section. Peat samples were dated and mangrove pollens and marine- and brackish-water diatoms were identified from collected samples. The eroded valley floor was, however, filled by marine sediments and subsequently with freshwater sediments due to the cyclic changes of transgression and regression. In Mekong River Delta the late Pleistocene-Holocene sequence since the last sea-level low stand is composed of an incised valley fill, comprising late to early Holocene fluvial to estuarine sediments, and Holocene deltaic sediment during the rise of sea-level around 8000–6000 years BP (Ta et al. 2001).The study has identified the magnitude of the mid-Holocene high stand of RSL in Bangladesh which is a unique finding of this study.

Due to these high stands, the coastline of the bay moved north. After 7500 years BP, the central part of Bangladesh was submerged by seawater until approximately 5500 years BP. Islam (2001) and Umitsu (1987, 1993) concluded that during the mid-Holocene, the coastline was slightly north of Khulna city and was continuously moving southward to the present sea level at 8000 years BP onwards. The present study strongly supports that the mid-Holocene paleo-coastline extended up to the central part of Bangladesh beyond the modern Padma-

Meghna course. The study also showed evidence of a higher RSL during the mid-Holocene. All the findings are new contributions in sea-level research, particularly for the Indo-Pacific Ocean region.

The research implies that the RSL in Bengal Lowland during the Holocene was influenced by some regional and local phenomena such as, ISM, climate change, delta building process, river system, and so on.

6.1 Recent Pattern and Future Projection of Sea Level Change in the World with Reference to Bangladesh

According to the world's leading scientists, sea level rise is "arguably one of the most important potential impacts of global climate change." Global average sea level has been rising over the last 100 years, and global warming is expected to increase the annual rate of sea level rise by two to five times. By the year 2100, sea level is projected to be approximately 47.5 cm (19 inches) higher than it is today (IPCC 1996). An increase of this magnitude could inundate coastal areas, erode beaches and exacerbate coastal flooding.

The greenhouse-gas-induced thermal expansion contribution to sea-level rise between 1880 and 1985 is estimated at 2–5 cm. Projections are made to the year 2025 for different forcing scenarios. For the period 1985–2025 the estimate of greenhouse-gas-induced warming is 0.6–1.0 °C. The concomitant oceanic thermal expansion would raise sea level by 4–8 cm (IPCC 1996).

Tide gauge data has been utilized by a number of researchers to calculate average rate of SLR (mm/year) since 1807–1980. At first, Gutenberg (1941) dealt with data ranging 1807–1937 and found a rate of 1.1 mm/year. Emery and Aubrey (1989) used monthly-mean sea-level data till 1982 at several tide-gauge stations along the Indian coast to estimate long-term trends in sea level. The trends showed considerable variability because even short records were included in the analysis. This inconsistency was also noted by Douglas (1991), leading him to conclude that the stations in the Indian subcontinent are not suitable for estimating global-mean sea-level rise. Using tide-gauge data all over the globe, Church et al. (2004) used reconstruction methods to determine the spatial pattern of sea-level variability during 1950–2000. These results were used to describe regional sea-level changes and suggest values close to 2.0 mm/year in the north Indian Ocean, except the northeastern part of the Bay of Bengal.

Unnikrishnan and Shankar (2007) showed statistically significant trends, obtained from records longer than 40 years yielded sea-level-rise estimates between 1.06 and 1.75 mm/year, with a regional average of 1.29 mm/year, when corrected for global isostatic adjustment (GIA) using model data. These estimates are consistent with the 1–2 mm/year global sea-level-rise estimates reported by the Intergovernmental Panel on Climate Change.

In Bangladesh, based on tide gauge data a very few researches have been done so far. For example, Das (1992) estimated a 25 cm rise of sea level during 1944–1964 (a rate of 1.25 cm/year) at Chittagong where as Kabir (1993) considered the fall to be 1.11 mm/year at Chittagong. Examination of relative sea level changes of Mongla (Khulna) by Kabir (1993) indicated a rise of 4.65 mm/year. Matin, (2009) mentioned a study by BWDB in the Gangetic Delta area the change in MSL is clearly positive rising by about 26 cm in the last 21 years at an annual average trend of 11.6 mm/year. SMRC study (2000) showed trends of sea level rise at three representative stations Cox's Bazaar, Char Changa and Hiron Point at 7.8, 6.0 and 4.0 mm/year respectively. The SMRC (2000) was based on 20 years (1977–1998) monthly tidal data.

A study conducted by Rashid and Mehruba (2011) deals with recent pattern of mean sea-level along the coast of Bay of Bengal. The MSL has been calculated on the hourly tidal data of three stations located in three different regions of the Bangladesh coast such as Hiron point located in the western coast in the Ganges delta (Sundarbans), Char Changa in the central coast and Cox's Bazar in the eastern coast of Bangladesh. The 10-years (1998–2008) data set was obtained from Bangladesh Inland Water Transport Authority (BIWTA). The tide gauges, Model no. MUNRO 1H109, U.K. are being used by BIWTA for data collection. The study provides MSL of selected stations of the coast of Bangladesh during 1998–2007 on the basis of hourly tidal data. This result indicates that the MSL of the three stations such as Hironpoint, Charchanga and Cox's bazaar are 1.89, 2.05, and 2.23 m respectively and rate of sea level rise have been found to be 4.32, 2.34 and 3.49 mm/year for Hironpoint, Charchanga and Cox's bazaar, respectively. The study found that during the monsoon season (June–Sep) sea level is much higher than the winter due to large scale rain fall within and outside the country.The study results support the global trend (0.3–3.3 mm/year). According to Emery (1980), long term record shows that mean annual sea-levels are rising in low and middle latitudes at a medium rate of 3 mm/year.

National Adaptation Programme for Action (NAPA 2005) team on the basis of third IPCC report and SAARC Meteorological Research Centre (SMRC) have established the likely climate change and SLR scenarios for Bangladesh (Table 6.1).

The study funded by the Asian Development Bank (ADB) said that IPCC's prediction (2007) about 0.18–0.59 m (Table 6.1) SLR did not take into account the sedimentation factors. One billion tones of sediment carried by Himalayan Rivers into Bangladesh every year, which are crucial in countering rises in sea levels. Sediments have been shaping Bangladesh's coast for thousands of years. Coastline, notably the Meghna estuary, which is one of the largest in the world, would rise at the same pace as the sea level growth. On the other hand, Chittagong coast is being uplifted since long past. Therefore, in order to get net sea level rise the estimated sea level rise obtained for the coast to be corrected by calculating the rate of vertical land movement, which are not available at present.

Most of the increase in global sea level is expected to come from thermal expansion of ocean water, followed by increased melting of glaciers and ice caps.

Table 6.1 Climate change scenarios for Bangladesh

Year	Sea level rise (cm)		
	Third IPCC (upper range)	SMRC	NAPA scenario
2030	14	18	14
2050	32	30	32
2100	88	60	88

(*Source* Modified from (Mohal et al. 2007)

A major source of uncertainty, however, concerns the polar ice sheets. Not only is there a lack of understanding of their current mass balance, but there is also considerable uncertainty regarding the ice sheets' possible dynamic response to climate change. These uncertainties, according to the United Nations' Intergovernmental Panel on Climate Change (IPCC), make a very large difference in the estimates of future sea level rise.

Just how much the polar region will contribute to future sea level rise is, as the IPCC indicates, a matter of great uncertainty. Ocean levels have always fluctuated with changes in global temperatures. During the ice ages, for example, when the Earth was 9 °F colder than today, much of the ocean's water was frozen in glaciers, and sea level was much lower than its current level. During the last interglacial period (120000 years ago) when the average temperature was 1.8–3.6 °F warmer than today, sea level was significantly higher (Titus and Narayanan 1995).

There is one problem with predicting future retreat or melting rate of the Greenland and Antarctic ice sheets: the ice sheets' current mass balance largely remains a mystery. Scientists do not yet understand how the ice sheets have contributed to historical sea level, so it is difficult to predict a future contribution with any certainty (IPCC 1996). Nevertheless, scientists do know that relatively small changes in the ice sheets could have a significant effect on sea level. The Antarctic ice sheet is particularly important, due to its sheer size—450 million cubic miles.

The study is an important contribution to the study of the sea level change in Bangladesh in the Holocene. The research reveals sea level variations relative to the present level caused by transgressions and regressions in the Bangladesh region. The results substantially contribute to the reconstruction of Holocene paleoenvironment of this area (Rashid 2011).

References

Banjeree PK (2000) Holocene and late pleistocene relative sea level fluctuations along the east coast of India. Mar Geol 167:243–260

Bookhagen B, Fleitmann D, Nishiizumi K, Strecker MR, Thiede RC (2006) Holocene monsoonal dynamics and fluvial terrace formation in the northwest Himalaya, India. Geology 34:601–604

Church JA, White NJ, Coleman R, Lambeck K, Mitrovica JX (2004) Estimates of the regional distribution of sea level rise over the 1950 to 2000 period. J Clim 17:2609–2625

Das SC (1992) Physical oceanography of the Bay of Bengal. In: Elahi KM, Sharif AHMR and Kalam AKMA (eds) Bangladesh, geography, environment and development. Bangaladesh National Geographical Association, Dhaka, p 36–52

Douglas, BC (1991) Global sea level rise. J Geophys Res 96(C4): 6981–6992

Emery KO (1980) Relative sea levels from tide-gauge records. Nat Acad Sci Washington D.C. 77(12):72–95

Emery KO, Aubrey DG (1989) Tide gauges of India. J Coast Res 5:489–501

Fleitmann D, Burns SJ, Mudelsee M, Neff U, Kramers J, Mangini A, Matter A (2003) Holocene forcing of the Indian monsoon recorded in a stalagmite from Southern Oman. Science 300:1737–1739

Gutenberg B (1941) Changes in sea level, postglacial uplift, and mobility of the Earth's interior. Bull Geol Soc Am 52:721–772

Hameed A, Achyuthan H, Sekhar B (2006) Radiocarbon dates and holocene sea-level change along the Cuddalore and Odinur Coast, Tamil Nadu. Curr Sci 91:362–367

Ingolfsson Olafur, Hjort Christian (1999) The Antarctic contribution to Holocene global sea level. Polar Res 18:323–330

Intergovernmental Panel on Climate Change (1996) Working group I report, The science of climate change, p 364–365

Isla FI (1989) Holocene sea-level fluctuation in the Southern hemisphere. Quatern Sci Rev 8:359–368

Islam MK (2001a) Evidences of neotectonics in and around Lalmai Hills, Bangladesh. Bangladesh J Geol 20:55–66

Islam MS (2001b) Sea-level changes in Bangladesh: the last ten thousand years. Asiatic aociety of Bangladesh, Dhaka

Kabir MH (1993) Secular variation in mean sea level along the coast of Bangladesh. Unpublished M. Sc. Thesis. University of Chittagong, Bangladesh

Long AJ (2001) Mid-Holocene sea level change and coastal evolution. Prog Phys Geogr 25:399–408

Matin MA (2009) Mean sea level variation in the Bangladesh coast. Unpublished article

Mohal N, Khan ZH, Rahman N (2007) Impact of sea level rise on coastal rivers of Bangladesh. IWM, Dhaka

Monsur MH (1995) An introduction to the quaternary geology of Bangladesh. City press and publications, Dhaka

NAPA (Bangladesh) (2005) Formulation of Bangladesh program of action for adaptation to climate change project

Rashid T (2011) Holocene sea level change in Bangladesh. In: Proceedings of the International Association for the Physical Sciences of Ocean (IAPSO), Australia, June 28-July 02

Rashid T and mehruba (2011) Recent pattern of relative sea level changes along the coast of Bay of bengal. Unpublished M. Sc. Thesis. University of Dhaka, Bangladesh

Rashid T, Monsur MH, Suzuki S, OOI N (2009) Reconstruction of Holocene paleoenvironment and evidence of sea-level changes in the Bengal lowland. J Geogr 118(4):631–645

SMRC (2000). The vulnerability assessment of the SAARC coastal region due to sea level rise:Bangladesh case. SMRC Publication, Dhaka (SMRC-No.3)

Ta TKO, Nguyen VL, Tateishi M, Kobayashi I, Saito Y (2001) Sedimentary facies, diatom and foraminifer assemblages in a late Pleistocene–Holocene incised-valley sequence from the Mekong River Delta, Bentre Province, Southern Vietnam: the BT2 core. J Asian Earth Sci 20:83–94

Titus James G, Narayanan Vijay K (1995) The probability of sea level rise. Environmental Protection Agency, U.S, p 138

Umitsu M (1987) Late quaternary sedimentary environment and landform evolution in the Bengal Lowland. Geog Rev Jpn 60:164–178

Umitsu M (1993) Late quaternary sedimentary environments and landforms in the Ganges Delta. Sed Geol 83:177–186

Unnikrishnan AS, Shankar D (2007) Are sea-level-rise trends along the coasts of north Indian Ocean coasts consistent with global estimates? Global Planet Change 57:301–307

Van Campo E, Cour P, Sixuan H (1996) Holocene environmental changes in Bangong Cobasin (Western Tibet), part 2: the pollen record. Palaeogeogr Palaeoclimatol Palaeoecol 120:49–63

Wang YJ, Cheng H, Edwards RL, He YQ, Kong XG, An ZS, Wu JY, Kelly MJ, Dykoski CA, Li XD (2005) The Holocene Asian Monsoon: links to solar changes and North Altantic climate. Science 308:854–857

Woodroffe SA, Horton BP (2005) Holocene sea-level changes in the Indo-Pacific. J Asian Earth Sci 25:29–43

MIX
Papier aus verantwortungsvollen Quellen
Paper from responsible sources
FSC® C105338

If you have any concerns about our products,
you can contact us on
ProductSafety@springernature.com

In case Publisher is established outside the EU,
the EU authorized representative is:
Springer Nature Customer Service Center GmbH
Europaplatz 3, 69115 Heidelberg, Germany

Printed by Libri Plureos GmbH
in Hamburg, Germany